D1714471

SEMICONDUCTOR TECHNOLOGY

SEMICONDUCTOR TECHNOLOGY

Processing and Novel Fabrication Techniques

Edited by

MIKHAIL E. LEVINSHTEIN

Ioffe Institute of Russian Academy of Science
St. Petersburg, Russia

MICHAEL S. SHUR

Rensselaer Polytechnic Institute
Troy, New York

A Wiley-Interscience Publication

JOHN WILEY & SONS, INC.

New York / Chichester / Weinheim / Brisbane / Singapore / Toronto

This text is printed on acid-free paper. ♾

Published simultaneously in Canada.

Library of Congress Cataloging in Publication Data:

Semiconductor technology : processing and novel fabrication techniques / edited by Mikhail Levinshtein, Michael Shur, translated by Minna Perelman
p. cm.
"A Wiley-Interscience publication."
Includes index.
ISBN 0-471-12792-2 (cloth : alk. paper)
1. Semiconductors. 2. Semiconductor doping. 3. Semiconductors--Defects. I. Levinshtein, M. E. (Mikhail Efimovich) II. Shur, Michael.
TK7871.85.S4453 1997
621.3815'2—dc21 97-1734
CIP

Printed in the United States of America

10 9 8 7 6 5 4 3 2 1

CONTENTS

PREFACE

The dawn of semiconductor research can be traced to the pioneering papers by Michael Faraday in the nineteenth century. However, the beginning of the modern semiconductor technology is much more recent, dating back to 1947 when John Bardeen, Walter Brattain, and William Shockley invented a bipolar junction transistor. This discovery initiated the development of many complicated and interrelated research and production techniques, which are now referred to as semiconductor technology.

This technology has had an enormous impact on modern civilization. However, we should point out that from 1947 to 1991, namely for more than 40 years, humankind had not one but two semiconductor technologies. Using the analogy with the Western and Eastern Roman empires, we may call these two technologies "Western" and "Eastern" technologies.

This is not to say that the device physics of a field effect transistor or of a Schottky diode was different in the East from that in the West. However, the ways to make a p-n junction, to fabricate ohmic contacts, or to passivate a semiconductor surface were often quite different in the East and in the West. Quoting Kipling, "There are nine and sixty ways of constructing tribal lays. And every of them is right!"

The Western scientists have been skeptical of the Eastern semiconductor technology. They have often considered the Soviet Union (or its successor, the Russian Federation, for that matter) as a backward country with an inferior infrastructure and a low standard of living. Quoting the Bible: "Can there any good thing come out of Nazareth?"

Yet one ought to remember that the Soviet Union had an enormous military-industrial complex. An important component of this complex was semiconductor and electronics science, which was heavily subsidized. It was this technology that was successful in supporting the Soviet military and space exploration programs.

Not surprisingly then, among many techniques, technological recipes, and characterization methods of the Eastern technology, can be found many excellent approaches, which can contribute enormously to the modern semiconductor technology. Nevertheless there is still a large number of these techniques that are practically unknown in the West, and this book is about bringing all of this information to the Western reader.

Semiconductor Technology will be useful to all scientists, engineers, technicians, and students interested in semiconductor materials or devices. It requires only a fairly general background in semiconductor physics and technology at the introductory textbook level.

We would like to thank the translator of this book, Mrs. Minna M. Perelman, for her help.

The editors and the authors would be very grateful for comments and suggestions, which can be submitted to one of the editors via e-mail: melev@nimis.ioffe.rssi.ru or shur@ecse.rpi.edu

MIKHAIL LEVINSHTEIN
MICHAEL SHUR
EDITORS

LIST OF SYMBOLS

$\vec{a}, a$	Period of diffraction grating and/or interference field
A_x	Atomic number of the nucleus
B_C	Height of the Coulomb barrier
C	Capacitance
ΔC	Capacitance change (variation)
d	Depth of the layer
d	Thickness
d_P	Thickness of the polymer diffusant film
D	Diffusivity
D	Irradiation dose
D_I	Diffusivity of self-interstitials
D_N	Diffusivity of electrically active aluminium atoms
D_V	Diffusivity of vacancies
D_I^*, D_V^*	Diffusivity of self-interstitials and vacancies under equilibrium conditions
E	Energy
E_g	Band gap
E_i	Level position in the forbidden gap
E_0	Initial particle energy
E_c	Conduction band edge
E_{ph}	Photon energy
E_{th}	Threshold of the nuclear reaction
E_v	Valence band edge
E_α	α-particle energy
E_β	β-particle energy
E_γ	γ-particle energy
f	Frequency

F	Electric field
F	Flux of the particles at a depth x
F_0	Flux of the particles at the surface
h	Depth of grating
$\langle hkl \rangle$	Crystallography direction
I	Current
I	Light intensity
I	Self-interstitial atom
I^*	Equilibrium concentration of self-interstitials
I_ν	Volume component of the reverse current
j_m	Bessel function of the first kind of the mth order
k	Boltzman constant
K	Compensation coefficient
K_0	Constant of bimolecular recombination
K_{tr}	Transmutation coefficient
e_i	Electron emission rate from deep level i to the conduction band
L_{Zn}	Solubility limit of Zn
m	Mass
m	Number of diffraction order
M	Doping impurity content in the polymer diffusant
M_a	Mass of the incoming particle
M_x	Mass of the nucleus
n	Carrier concentration
n	Refractive index
N	Concentration
N_{Au}	Gold bulk concentration in silicon
N_d	Shallow donor concentration
N_{dis}	Dislocation density
N_{imp}	Concentration of introduced impurity atoms
N_{In}	Indium concentration in the layer
N_o	Concentration of target nucleus
N_s	Impurity surface concentration
N_{ss}	Surface state density
N_{Zn}	Zinc concentration introduced into the near-surface region
q	Charge of the electron
r	Reflectivity
r	Radius of the nucleus

S	Surface recombination velocity
Si_{ξ}	Substitutional silicon atom
t	Time
t	Duration of irradiation
T	Temperature
T_S	Substrate temperature
U	Potential
U	Voltage
U_{Br}	Breakdown voltage
U_{Op}	Operation voltage
V	Lattice vacancy
V^*	Equilibrium concentration of vacancies
V_d	Velocity drift
W	Wafer thickness
x_{Bi}	Mole fraction of bismuth in the solvent
x_j	p-n junction depth
Z_a	Charge of the incoming particle
Z_x	Charge of the nucleus
α	Angle of light incidence
Γ	Argument of the Bessel function
ε	Dielectric permittivity
η	Diffraction efficiency
λ	Light wavelength
λ	Decay constant of radionuclide
μ	Carrier mobility
ν	Number of nuclear interactions
ρ	Resistivity
ρ_0	Initial resistivity
σ	Nuclear interaction crossection
σ_s	Surface conductivity
τ	Minority carrier lifetime
τ_0, τ_∞	Minority carrier lifetime at low and high injection level, accordingly
τ_n, τ_p	Electron and hole lifetime, accordingly
ω	Angular frequency

CONTRIBUTORS

Lydvig V. Belyakov, The Ioffe Institute, Polytechnicheskaya 26, 194021, St. Petersburg, Russia.
Vladimir V. Chaldyshev, The Ioffe Institute, Polytechnicheskaya 26, 194021, St. Petersburg, Russia.
Aleksei T. Gorelonok, The Ioffe Institute, Polytechnicheskaya 26, 194021, St. Petersburg, Russia.
Dmitri N. Goryachev, The Ioffe Institute, Polytechnicheskaya 26, 194021, St. Petersburg, Russia.
Elena G. Guk, The Ioffe Institute, Polytechnicheskaya 26, 194021, St. Petersburg, Russia.
Natal'ya D. Ilyinskaya, The Ioffe Institute, Polytechnicheskaya 26, 194021, St. Petersburg, Russia.
Alexandr V. Kamanin, The Ioffe Institute, Polytechnicheskaya 26, 194021, St. Petersburg, Russia.
Vitalii V. Kozlovskii, Experimental Physics Department, St. Petersburg Technical University, 195251, St. Petersburg, Russia.
Irina A. Mokina, The Ioffe Institute, Polytechnicheskaya 26, 194021, St. Petersburg, Russia.
Sergey V. Novikov, The Ioffe Institute, Polytechnicheskaya 26, 194021, St. Petersburg, Russia.
Natal'ya M. Shmidt, The Ioffe Institute, Polytechnicheskaya 26, 194021, St. Petersburg, Russia.
Valentina B. Shuman, The Ioffe Institute, Polytechnicheskaya 26, 194021, St. Petersburg, Russia.
Nikolai A. Sobolev, The Ioffe Institute, Polytechnicheskaya 26, 194021, St. Petersburg, Russia.

Ol'ga M. Sreseli, The Ioffe Institute, Polytechnicheskaya 26, 194021, St. Petersburg, Russia.
Tat'yana A. Yurre, Technological Institute, Moskovskii 26, 198013, St. Petersburg, Russia.
Leonid F. Zakharenkov, Experimental Physics Department, St. Petersburg Technical University, 195251, St. Petersburg, Russia.

SEMICONDUCTOR TECHNOLOGY

CHAPTER 1

INTRODUCTION TO SEMICONDUCTOR TECHNOLOGY

E. G. GUK and N. M. SHMIDT

The technology of semiconductor devices involves many steps, beginning with growing single crystals and ending with the packaging device and assembly. There are many excellent monographs that describe in detail practically all such aspects of the process, including relevant physical and chemical characteristics and technological advances. The best among such monographs are devoted to the processes of diffusion (Sze 1983), epitaxy (Sze 1983), and ion implantation (Russel et al. 1978). Other books contain full descriptions of the manufacturing technologies of certain devices (Burger and Donovan 1967; Ghandhi 1994).

This book considers a few new technological methods that have been developed in recent years. The information provided here has never been described in a book in a systematic fashion. Previously only a few original articles containing limited information were available to Western readers. This book therefore should appeal to technicians and engineers who design semiconductor devices and are interested in promising new technological methods. The book describes doping, gettering, passivation, and etching of semiconductors.

Semiconductor Technology: Processing and Novel Fabrication Techniques,
Edited by M. Levinshtein and M. Shur.
ISBN 0-471-12792-2

1.1. DOPING

Doping or the introduction of impurities into a material, is widely used in the technology of semiconductors and semiconductor devices. Doping makes it possible to solve the following problems:

1. To obtain the material with a certain concentration of majority carriers due to the introduction of electrically active impurities.
2. To obtain semi-insulating materials ($\rho > 10^5\ \Omega \cdot \text{cm}$) and materials with a controlled reduction of the minority carrier lifetime due to the introduction of impurities that form deep levels in the semiconductor's band gap.
3. To improve structural perfection of semiconductor materials due to a reduced concentration of the crystal lattice defects. (As this occurs, the mobility of the majority carriers and the lifetime of the minority carriers usually increases; this result is achieved by introducing isovalent impurities, namely impurities of the same group of periodic system as the semiconductor crystal atoms.)

Clearly the term "doping" embraces different processes. A standard practice is to distinguish between purely metallurgical doping, which introduces the impurity in the growing of a single crystal or a semiconductor layer, and the postgrowth doping, which introduces the impurity into a full-grown semiconductor wafer, into the epitaxial layer, and device region. The postgrowth doping comprises processes of diffusion, ion implantation, and transmutation doping. Chapter 2 deals with a relatively new method of transmutation doping by charged particles. In Chapter 3 the use of the principally new sources of diffusion (the polymer diffusants) in the traditional process of diffusion is discussed. Chapters 4 and 6 elucidate the peculiarities of doping by rare-earth elements and isovalent impurities in the growing of epitaxial layers and semiconductor single crystals.

1.1.1. Transmutation Doping

Transmutation doping is principally different from the standard doping in that the doping impurities are not externally introduced into the semiconductor. They are rather formed in the process of irradiating the semiconductor by protons, alpha-particles, neutrons, and γ-quanta directly from the atoms of the initial material as a result of the adequate nuclear reactions (Smirnov 1981). Doping by neutrons has been investigated for a long time, and it is employed in industry to obtain high-resistivity uniformly silicon ingots. Currently transmutation doping by charged particles has become a topic of interest. It has all the advantages of the neutron doping and is a technique of choice for providing controlled locality in depths ranging from 0.1 μm to 1 mm. It also enables local doping on the wafer by means of photolithography for device fabrication. Transmutation doping by charged particles offers a unique possibility of forming deep channels. The width of these channels is primarily determined by the mask resolution and can be in the micrometer range. This technology allows one to fabricate power field effect transistors with a vertical channel (Kozlovskii et al. 1992).

The experimental results of this rapidly developing technology have not yet been systematized in monographs and reviews. Chapter 2 is written by two leading Russian experts in the field of the radiation doping of semiconductors. They consider the physical basis of transmutation doping by charged particles, and they review the effects for semiconductors (Si, SiC, III-V, and II-IV compounds) and other materials (high-temperature superconductors, ferroelectrics). They also consider the applications of transmutation doping for devices fabrication. The advantages of transmutation doping have been clearly demonstrated by doping GaAs with donor impurities to the depth of 100 μm (Afonin et al. 1988) and by doping SiC with acceptor impurities to the depth of hundreds of micrometers (Kozlovskii et al. 1992). In the first case a prolonged and complicated process of forming LPE high-voltage pulse diodes using low-doped ($\sim 10^{15}$ cm^{-3}) GaAs has been replaced by an hour-long process of transmutation doping. In the second case, a high-temperature process (> 2500°C) of Al diffusion into SiC, lasting for many hours, used in the fabrication of light-emitting diodes, has been replaced by an hour-long proton irradiation with the subsequent hour-long annealing at the temperatures not higher than 1000°C (Didik et al. 1991).

1.1.2. Diffusion

Diffusion has been used in semiconductor technology since the 1950s. It consists in introducing doping impurities at a high temperature into the surface layer of a semiconductor from a gas phase or from solid sources, spinned on the semiconductor surface. The diffusion temperature does not exceed the melting temperature. There are many methods of carrying out the process of diffusion (Burger and Donovan 1967), many of which have been developed high perfection. However, the rapid development of microelectronics, optoelectronics, and nanoelectronics imposes more rigid requirements on the diffusion impurity and reproducibility.

The reproducible formation of diffusion layers with a preassigned exact surface concentration of impurities, retaining a high degree of uniformity on large wafers with diameters ≥ 100 mm, still remains a challenge for silicon planar technology, which uses diffusion as one of the basic processes. One possible solution of the problem is considered in Chapter 3. The new types of diffusion sources, developed by the authors, are based on film-forming polymer layers. These layers have indubitable advantages over conventional diffusion sources. The use of polymer spin-on films enables one to obtain the silicon avalanche diodes with work area over 1 cm^2 with practically homogeneous avalanche breakdown over the entire area (Zubrilov and Shuman 1987; Zubrilov et al. 1989). The area of the homogeneous avalanche breakdown can be increased by an order of magnitude without reducing the quality of the device. This result has been achieved because the atoms of the doping impurity are directly introduced into the three-dimensional polymer network and become chemically bounded with the polymer molecules and uniformly distributed within the network. The authors managed to achieve a uniform and reproducible introduction of impurities into silicon. Their results are as good as those obtained using a gas phase or ion implantation, and the method is both simpler and cheaper. Besides, the possibility of introducing any required element or several elements into diffusant makes it possible to conduct a simultaneous diffusion of shallow and deep impurities and control the concentration of these impurities in the diffusion layer (Guk et al. 1986).

The use of new diffusants is also promising in the technology of III-V compounds. The process of diffusion in III-V compounds technology has not received such wide acceptance as in silicon technology because the temperature at which the decomposition of III-V semiconductors begins is often lower than that required for the diffusion. In the process of diffusion it is also necessary to maintain high vapor pressure of the column V component (Van Gurp 1987).

Beginning in the 1980s, the process of diffusion from spin-on films has been gaining acceptance in the III-V compounds technology (Arnold et al. 1984). The process is carried out in an open tube filled with an inert gas. The silicon organic film, containing the doping impurity spinned-on by centrifuging on the semiconductor surface, serves as the source. This spinned-on film prevents the decomposition of the semiconductor surface in the diffusion process. The main drawback of this method is a poor control of the mechanical stress appearing as the film of the diffusant is being formed. The new diffusants, described in Chapter 3, allow one to overcome this drawback, and they offer a new direction for III-V compound planar technology.

Zn diffusion from the polymer diffusants into InP, InGaAs, InGaAsP, described in Chapter 3, shows that it is possible to retain the structure perfection of a semiconductor in the diffusion process and to control the concentration and the depth of penetration of the impurity by changing of the doping impurity content in the polymer diffusant. This method proves to be as simple and practical as those used in the silicon standard technology (Belyakov et al. 1992).

1.1.3. Isovalent Doping

Isovalent doping is doping with impurities belonging to the same periodic group as the semiconductor atoms, which they substitute. Such a doping creates no additional impurity levels in the band gap. However, it may lead to changes in the length of the bonds and in the geometry of the intrinsic point defects of the lattice. This affects the formation of the effective centers of the generation and recombination of minority carriers and makes it possible to control carrier mobility and the minority carrier lifetime in semiconductors. The effect of isovalent doping may even lead to the redistribution of the amphoteric impurity between the sublattices in the direct gap III-V semiconductors and, consequently, to the change in the conductivity type of the material.

The first investigations of the behavior of isovalent impurities in the semiconductors were started 30 years ago (Gross and Nedzevetskii 1963). This work made it possible to fabricate highly efficient light-emitting diodes based on the indirect semiconductor GaP : N (Bergh and Dean 1976). In the early 1980s the isovalent doping received much attention because dislocation-free bulk crystals and epitaxial films obtained had a low concentration of defects. The main attention was focused on investigating GaAs and other direct gap III-V compound semiconductors.

Chapter 6 reviews the experimental results on isovalent doping of direct gap compounds III-V grown by different epitaxial methods (vapor-phase, molecular-beam, and liquid-phase epitaxy) and obtained by ion implantation. The main principles and mechanisms of isovalent doping of III-V compounds are considered there. The authors show that it is possible to have a subtle control of the material properties for all main methods of epitaxial growth. For example, the isovalent doping with indium allows one to reduce the defect concentration in the GaAs layers, grown by vapor-phase and molecular-beam epitaxy by more than an order of magnitude.

In ion implantation the isovalent impurity can be used simultaneously with the donor impurity. This increases the donor activation efficiency and thus the electron concentration in the layer. Chapter 6 describes isovalent doping of GaAs, InP, GaSb, and other semiconductor layers with the emphases on general trends typical for this technology.

1.1.4. Rare-Earth-Element Doping

Rare-earth elements are the most chemically active elements of the lantanoid group. Their activity determines their application in the semiconductor technology. The whole spectrum of these applications has not yet been studied, and the properties of the semiconductors with the rare-earth elements are yet to be established.

In semiconductor technology the rare-earth elements are used to facilitate the growth of the high purity layers and ingots of III-V compounds with high electron mobility and for highly efficient luminescence devices based on intracenter f-f luminescence in a semiconductor matrix of Yb^{3+} ions (Zakharenkov et al. 1981) and Er^{3+} (Ennen et al. 1983) in the spectral region of 1.0 and 1.54 μm.

Chapter 4 describes the effect of rare-earth elements on the electrophysical parameters of the InP, GaAs bulk single crystals, and the InP, InGaAs, InGaAsP epitaxial layers and relevant technological methods. In particular, it considers the technological aspects of synthesis and growing ingots of semi-insulating GaAs and InP crystals.

The authors succeeded in solving a fundamental technological problem of creation of a stable protective coating made of oxides of rare-earth elements on the quartz equipment. Such coating prevents the interaction between the melt and the container material up to temperatures of the order of 1000°C. As a result of those investigations, a promising reproducible process of using the rare-earth elements in the synthesis of various III-V compounds has been developed (Zakharenkov 1995).

Rare-earth elements interacting in the melt with the background impurities act as a getter and purify epilayers or monocrystals during growth. The parameters of monocrystals and epilayers grown with rare-earth elements are as good as those obtained by more complicated and expensive techniques (MBE, MOVPE). The authors have obtained monocrystals of semi-insulating GaAs with resistivity exceeding $10^8 \Omega \cdot$ cm and electron mobility larger than 6000 $cm^2 \cdot V^{-1} \cdot s^{-1}$ at 300 K; the InP and InGaAs epilayers with the electron concentration $(1-2) \cdot 10^{13}$ cm^{-3} and the electron mobilities of 6500 and 100,000 $cm^2 \cdot V^{-1} \cdot s^{-1}$ at 300 and 77 K, respectively. Rare-earth element technology allows one to fabricate III-V devices that are as good as the best devices fabricated by other techniques (Tsang 1985).

1.2. GETTERING

Gettering removes undesirable impurities and defects from the active regions of semiconductor devices; it is one of the key processes in the semiconductor technology. In the early stages of semiconductor technology the term "gettering" related to methods that prevent unintended impurities from getting into the growing crystal or into a semiconductor device during fabrication. Called *external gettering*, the direct interaction of the getter with the unintended impurity requires the removal of the getter saturated with the impurity. The typical example involves the gettering of the melt in the liquid-phase epitaxy process.

The properties of semiconductors may be affected by the intrinsic point defects as much as by unintentional impurities. In the 1970s it was established that when the concentration of unintentional impurities in silicon was less than 10^{13} cm^{-3}, the generation and relaxation of nonequilibrium intrinsic point defects became a dominant factor. These processes determined the electrical activity of the generation-recombination centers. The techniques that control the generation and relaxation of the intrinsic point defects and use these control to neutralize the unintended impurities or to remove them from the active regions are referred to as *intrinsic gettering*. Since the generation-relaxation of the point defects takes place at every stage of the device fabrication, intrinsic gettering has become a powerful tool that allows one to achieve control over the properties of semiconductors.

Intrinsic gettering is achieved by the following methods:

1. Determine the sources or sinks of the intrinsic defects of the lattice outside of the semiconductor active region. This is usually done by forming a ruggy layer on the backside of the semiconductor substrate. This layer can be formed by a laser beam (Perce and Zaleckas 1979), by the implantation of impurities with different tetrahedral radii (Bronner and Plummer 1985), by applying layers of porous silicon (Chen and Silvestry 1982) or metal silicides (Murarka 1986).
2. Choose conditions that will relax the intrinsic defects inside the active layer and allow the adsorption of the unintended impurities by the sinks of nonequilibrium intrinsic point defects within the device. Such conditions include the temperature, the duration of thermal treatment, and mechanical stresses, either existing in the semiconductor or specifically created for this purpose (Hu 1980).

3. Choose a sequence of technological operations that will annihilate the defects introduced in a given technological step with the defects-antipodes introduced in the next technological operation. For instance, the thermal oxidation of silicon is accompanied by the generation of interstitial Si atoms at the $Si-SiO_2$ interface. The subsequent diffusion of boron is accompanied by the generation of vacancies. If the heat-treatment regime is chosen correctly, one will be able to retain the structural perfection of the semiconductor in the active region of the p-n junction and reduce the leakage current by several orders of magnitude (Ivanov et al. 1982).

A great number of gettering methods have been developed in silicon technology, and the attained results are often remarkable. The concentration of unintentional impurities and electrically active centers can be reduced below 10^{11} cm^{-3}. However, the modern technology has more rigid requirements that demand perfection of the gettering methods. Any further progress will be only achieved by a deeper understanding of the generation and relaxation of nonequilibrium intrinsic defects in semiconductors.

The methods of intrinsic gettering in the technology of III-V compounds have not yet been developed to the same extent as in silicon technology. The reason is a lower level of the III-V compound semiconductor technology, namely higher contamination by unintended impurities and unoptimized cleaning methods. The multicomponent composition of these semiconductors and the volatility of the group V component give rise to a large variety of nonequilibrium intrinsic point defects which hinders their identification and the development of the gettering methods. Besides, these materials are more plastic, and hence the conditions of conducting the gettering experiments must be observed more strictly. The supersaturation of the lattice with the intrinsic nonequilibrium defects can be eliminated by the local change of the composition, which is very hard to detect or control. All these factors suggest that for the effective control of the generation and relaxation of the nonequilibrium intrinsic defects, it is necessary to acquire a better understanding of the processes by looking more closely at the effective centers of generation-recombination in these materials.

This book does not have a separate chapter devoted to the process of gettering, but Chapters 3 to 5 contain original methods of external and intrinsic gettering in different semiconductors. Effective methods of external gettering of the background impurities in the III-V component technology are described in Chapter 4. The authors show that the introduction of rare-earth elements into the solution during the InP epilayers and the InGaAsP and InGaAs compounds growth reduces the level of contaminations by background impurities (S, O, Ge, Si) by several orders of magnitude (Gorelonok et al. 1988).

Chapter 3 demonstrates how new polymer diffusants can be used for gettering unintended impurities in the fabrication of polycrystal and single-crystal silicon solar cells. The developed methods make it possible to increase the minority carrier lifetime in a polycrystal silicon to 10–15 μs compared to 3–5 μs obtained by standard methods (Guk et al. 1990).

Chapter 5 reviews the generation-relaxation of nonequilibrium intrinsic defects in Si and some new suggestions on the role of these defects in the formation of the effective centers of generation-recombination in silicon. For example, Vyzchigin et al. (1991) have observed deep level centers whose formation is linked to intrinsic defects. Advantages of gettering are demonstrated by considering the technology of high-voltage silicon devices. Gettering allows one to fabricate diodes with a breakdown voltage of 3 to 6 kV and with the current of 1250 A. For the diodes in the 23-cm^2 work area, the homogeneous avalanche gain was observed at the reverse current density of 1 A/cm^2 practically over the entire diode area (Vil'yanov et al. 1989).

1.3. SURFACE PASSIVATION

The main goal of passivation is to minimize the influence of the semiconductor surface on the bulk properties of the semiconductor. To illustrate the importance and difficulty of passivation the technology of forming the $Si-SiO_2$ interface (Sze 1983), we recall the words of the great Swiss physicist Wolfgang Pauli who remarked that God created the volume but the devil made the surface. Passivation technology has progressed a long way over the last 40 years. It has developed along with the studies of the physical and chemical properties of the Si–SiO_2 interface, the conditions of its formation, and the diagnostic methods. The simultaneous use of passivation and gettering methods has reduced the surface state density 10,000 times—from 10^{13} to $10^9\,eV^{-1} \cdot cm^{-2}$ (Sugano 1980) and the surface recombination velocity more than a thousand times —from $10^5\,cm \cdot s^{-1}$ to several tens $cm \cdot s^{-1}$ (Ivanov et al. 1982). It has become possible to retain the stability of interface parameters for many years.

Despite this success there is no clear knowledge of the interactions among physical, chemical, structural, and electronic properties of the Si–SiO_2 interface and the conditions of its formation. No general principles of the interface formation have been fully devised, except for certain cases largely depending on the unique properties of the intrinsic silicon oxide.

The lack of such general understanding makes it difficult to develop the surface passivation processes for other semiconductors, specifically for III-V compounds. The problem of III-V compounds passivation is made even more complicated by their specific features, such as a complex phase composition of native oxides, including the thermodynamic instable semiconducting and conducting phases, by the Fermi level pinning at the semiconductor surface, by their greater plasticity compared to silicon, and by an intensive interaction with oxygen.

Nevertheless, some definite progress has been achieved in the passivation of III-V compounds. The best results were obtained for the InP-dielectric interface (Post et al. 1983). Different authors have used SiO_2, Si_3N_4, and two versions of a two-layer dielectric: SiO_2/a thin layer of the native oxide and Al_2O_3/a thin layer of the native oxide. The main achievements were the reduction of the surface states density to the values less than $10^{11}\,eV^{-1} \cdot cm^{-2}$, the fixed charge to the values on the order of $5 \cdot 10^{10}\,cm^{-2}$, and the hysteresis of the *C-V* characteristics to approximately 0.5 V. However, the main problem—the time stability of the interface parameters—remains unsolved. Still the passivation of InP and InGaAs *p-n* junctions peripheries provides for a long-term operation of photoreceivers based on these materials (Chane 1986). Chapter 7 contains the description of the passivation processes for InP and InGaAs *p-n* junction periphery with polyimide films, providing the retention of the sufficiently small densities of the dark current of

$10^{-7} A \cdot cm^{-2}$ and the surface recombination velocity of $\leq 10^3$ $cm \cdot s^{-1}$ with lifetime over 500 hours at 80°C. This process is simpler and cheaper than standard passivation. The results open up new possibilities for fabricating InP and InGaAs *p-n* junctions.

Chapter 7 also gives the results of the investigations of the interface InP—the InP oxide, formed by plasma oxidation. It shows that by changing the phase composition, one can effectively control the interface parameters and obtain a small density of surface states $-5 \cdot 10^{10}$ $eV^{-1} \cdot cm^{-2}$ and the hysteresis of *C-V* characteristics less than 0.2 V. Nevertheless, the achieved interface time stability, though comparable to the best cited in publications (Tardy et al. 1991), is still very far from that obtained for the Si–SiO_2 interface. The chapter authors reach an interesting conclusion that there is a certain link between the time instability of interface of III-V compound–dielectric parameters and structural peculiarities of the dielectric formed by low-temperature methods.

1.4. ETCHING

Etching is used for different purposes ranging from the removal of the ruptured semiconductor layer after mechanical treatment to the incipient technology of forming porous silicon possessing nontrivial and promising properties. The term "etching" has recently become quite polysemantic and now embraces such dissimilar processes as ionic, plasma, chemical (electrochemical), and other types of etching.

We will not dwell on the confusion of terms pertaining to the concepts of chemical and electrochemical etching. We would like to note that both processes are of the electrochemical nature and are redox processes with the subsequent dissolution of the oxidation products. A fairly wide spectrum of multicomponent etchants has been developed for different semiconductor materials (Sangwal 1987). It is well-known that the rate of etching depends on the chemical composition and temperature of the etchant and on the crystallographic orientation of the semiconductor surface. Photochemical etching, where the etching rate depends on the illumination of a surface, is used less frequently.

Chapter 8 describes a very interesting and nontraditional technique of photochemical etching. The authors discuss in detail how to obtain holographic diffraction grating by means of maskless chemical etching. The first data on conducting such etching on silicon were published in 1970 (Dalisa et al. 1970). A more elaborate development of the process and creation of certain technology have been performed by authors of Chapter 8 since 1974. A detailed theory of photochemical etching has been developed, and requirements to the adequate technological processes have been formulated.

This technology was applied to the fabrication of new optoelectronic devices and to the investigation of the physical parameters of semiconductors. In a comparatively short period of time, diffraction gratings were fabricated on more than 20 semiconductor materials, single and polycrystalline, amorphous, and so on. Using this gratings, some of the first world semiconductor lasers with light beams that have small angular divergence and distributed feedback were fabricated. A large number of experiments dealing with the excitation of surface polaritons in semiconductor-metal structures have been performed. Chapter 8 also describes the pioneer work on manufacturing holographic diffraction grating on the surface of metals using photochemical etching.

The unique features of this technology were demonstrated forming the sinusoidal grating, which cannot be achieved by other methods. Photochemical etching has the best resolution and does not violate the crystal structure of the materials.

1.5. CONCLUSION

This book describes the basic technological processes of semiconductor device fabrication. No electrical engineer, no matter what material and with what devices he or she is working with, can do without at least one or more processes described in this book. For any of these processes, the book describes new original approaches and techniques. The aim of this book is to draw the attention of electrical engineers and designers of the semiconductor devices to new promising technological methods.

REFERENCES

Afonin, O. F., Viktorov, B. V., Zabrodin, B. V., et al. (1988). *Sov. Phys. Semicond.* 22, 35–38.

Arnold, N., Schmitt, R., and Heime, K. (1984). *J. Phys. D, Appl. Phys.*, 17, 443–474.

Belyakov, S. V., Busygina, L. A., Gorelenok, A. T., et al. (1992). *Sov. Tech. Phys. Lett.* 18, 415–416.

Bergh, A. A. and Dean, P. J. (1976). In *Light-Emitting Diodes*. Clarendon Press, Oxford.

Bronner, G. B. and Plummer, J. D. (1985). *Appl. Phys. Lett.* 46, 510–512.

Burger, R. M. and Donovan, R. P. (1967). In *Fundamentals of Silicon Integrated Device Technology: Oxidation, Diffusion and Epitaxy*. Prentice-Hall, Englewood Cliffs, NJ.

Chané, J. P., Martin, B. G., Patillon, J. N., and Gentner, J. L. (1986). *Le vide les couches minces* 41, 203–206.

Chen, M. C. and Silvestri, V. I. (1982). *J. Electrochem. Soc.* 129, 1294–1299.

Dalisa, A. V., Zwicker, W. K., Debitetto, D. J., and Harnack, P. (1970). *Appl. Phys. Lett.* 17, 208–210.

Didik, V. A., Kozlovskii, V. V., Malkovich, R. Sh., et al. (1990). In *Proc. Int. Conf. on Radiation Materials Science* 6, (Alushta, Crimea) pp. 50–53 (in Russian).

Ennen, H., Schneider, J., Pomrenke, G., and Axmann, A. (1983). *Appl. Phys. Lett.* 43, 943–945.

Ghandhi, S. K. (1994). In *VLSI Fabrication Principles: Silicon and Gallium Arsenide*. Wiley, New York.

Gorelenok, A. T., Gruzdov, V. G., Kumar, R., Mamutin, V. V., et al. (1988). *Sov. Phys. Semicond.* 22, 21–26.

Gross, E. F. and Nedzevetskii, D. S. (1963). *DAN SSSR* 152, 309–312.

Guk, E. G., Shuman, V. B., Tarkhin, D. V., et al. (1990). *Proc. 2nd Int. Conf. on Polycrystalline Semiconductors* 11, pp. 255–258.

Hu, S. M. (1980). *J. Appl. Phys.* 51, 3666–3671.

Ishii, H., Hasegava, H., Ishii, A., and Ohno, H. (1988). *Surf. Sci.* 41/42, 390–394.

Ivanov, E. I., Lopatina, L. B., Sukhanov, V. L., et al. (1982). *Sov. Phys. Semicond.* 16, 129–132.

Kozlovskii, V. V., Zakharenkov, L. F., and Shustrov, V. A. (1992). *Sov. Phys. Semicond.* 26, 1–11.

Murarka, S. P. (1983). In *Silicides for VLSI Applications*. Academic Press, New York.

Ohno, T. and Shiraishi, K. (1990). *Phys. Rev. B.*, 42, 11194–11197.

Pearce, C. W. and Zaleckas, V. J. (1979). *J. Electrochem. Soc.* 126, 1436–1441.

Russel, H., Ruge, J., and Teubner, B. G. (1978). In *Ionenimplantation.* Stuttgart.

Sandroff, S. J., Nottenburg, R. N., Bischoff, J. C., and Bhat, R. (1987). *Appl. Phys. Lett.* 51, 33–37.

Sangwal, K. (1987). In *Etching of Crystals: Theory, Experiment and Application*. North-Holland, Amsterdam.

Smirnov, L. S. (1981). In *Doping of Semiconductors by the Nuclear Reaction Method.* Novisibirsk (in Russian).

Sugano, T. (1980). *Surf. Sci.* 98, 145–153.

Sze, S. M. (1983). In *VLSI Technology*. McGraw-Hill, New York.

Tardy, J., Thomas, T., and Viktorovich, P. (1991). *Appl. Surf. Sci.* 50, 383–387.

Tsang, W. T. (1985). In *Lightwave Communications Technology Part D Photodetectors*. Academic Press, Orlando.

Van Gurp, G. J. (1987). In *GaAs and Related Compounds*. Inst. Phys. Conf. Ser. 91, pp. 509–512.

Vil'yanov, A. F., Vyzhigin, Yu. V., Gresserov, B. N., et al. (1989). *Sov. Phys. Tech. Phys.* 34, 1184–1185.

Vyzhigin, Yu. V., Sobolev, N. A., Gresserov, B. N., and Sheck, E. I. (1991). *Sov. Phys. Semicond.* 25, 799–803.

Zakharenkov, L. F., Kasatkin, V. A., Kesamanly, F. P., and Sokolova, M. A. (1981). *Sov. Phys. Semicond.* 15, 1614–1615.

Zakharenkov, L. (1995). *Microelectr. J.* 26, 55–67.

Zubrilov, A. S. and Shuman, V. B. (1987). *Sov. Phys. Techn. Phys.* 32, 1105–1106.

Zubrilov, A. S., Kotin, O. A., and Shuman, V. B. (1989). *Sov. Phys. Semicond.* 23, 380–382.

CHAPTER 2

TRANSMUTATION DOPING OF SEMICONDUCTORS BY CHARGED PARTICLES

V. V. KOZLOVSKII and L. F. ZAKHARENKOV

The growing complexity of semiconductor electronics has revealed the limitations of present-day methods of doping: thermal diffusion and ion implantation. Most of the investigators from the United States, Japan, and Western Europe have considered mainly modifying the techniques of ion implantation and epitaxy, in particular the molecular-beam-epitaxy (MBE) technique. In contrast, the Russian work on semiconductor doping has emphasized the development of the radiation methods of doping.

At the present time there are four basic methods of radiation doping of semiconductors (Table 2.1) (Kozlovskii 1992):

- Doping of semiconductors by radiation defects, which appear under the irradiation of a semiconductor by neutrons, protons, electrons, or γ rays.
- Ion-enhanced processes comprising the recoil implantation processes and ion beam mixing.
- Radiation-enchanced diffusion.
- Transmutation nuclear doping.

Semiconductor Technology: Processing and Novel Fabrication Techniques,
Edited by M. Levinshtein and M. Shur.
ISBN 0-471-12792-2

TABLE 2.1

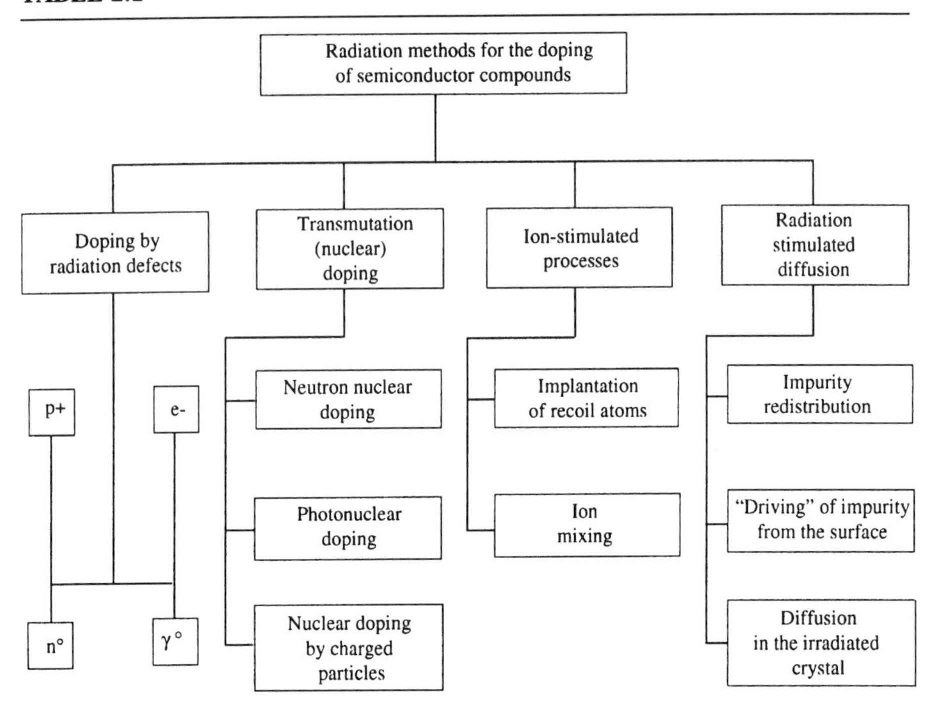

The first three methods of doping have been considered in detail in several reviews (e.g., Smirnov 1980; Vavilov et al. 1988). The present chapter is devoted to the fourth method—the transmutation nuclear doping.

The method of transmutation doping was proposed in the early 1950s (Lark-Horowitz et al. 1948; Lark-Horowitz 1950) and became widely used in the 1960s and 1970s for the uniform doping of silicon ingots. Several books and articles (Meese 1979; Smirnov 1981) emphasized doping processes based on nuclear reactions with thermal neutrons ("neutron" doping) and γ rays ("photonuclear" doping). The use of nuclear reactions involving charged particles was hardly even considered for transmutation doping: Only three original experimental investigations on this subject can be mentioned:

Lark-Horowitz et al. (1948), Trey and Oberhauser (1957), Dolgolenko and Shakhovtsev (1970). The new method of doping could not be used in those years because the isotope sources did not provide sufficiently intensive particle flux and did not allow for change in the energy of particles.

Toward the end of the 1970s, there was renewed interest in transmutation doping due to the development of the acceleration techniques and intensive particle fluxes with a wide range of energies. Modern accelerators use either linear or cyclic methods of acceleration (Komar 1975; Goldin 1983). At the present time, transmutation doping is conducted primarily in cyclic accelerators. At the Laboratory of Radiation Methods in St. Petersburg, Russia, systematic work on simulating transmutation doping processes began in the early 1980s, and subsequent experimental research started in 1985 using the cyclotron NGC-20 has been put into operation.

This chapter contains a review of both experimental and theoretic studies on transmutation doping by charged particles carried out during the last decade. We also consider applications of transmutation doping in semiconductor device technology.

2.1. NUCLEAR REACTIONS INVOLVING CHARGED PARTICLES

The peculiarity of nuclear reactions involving charged particles is related to the Coulomb repulsion of an incoming particle and the semiconductor matrix nucleus. Besides, in case if another charged particle escapes from the excited compound nucleus, that particle must also obtain the energy sufficient to overcome the Coulomb barrier. That energy has to be obtained from the compound's nucleus. Thus, when particle a interacts with nucleus X, the height of the Coulomb barrier is

$$B_{\mathrm{C}} = \frac{Z_x Z_a e^2}{r_x} \simeq \frac{Z_x Z_a}{A_x^{1/3}}, \tag{2.1}$$

where r_x is the radius of the nucleus; Z_x, Z_a is the charge of the

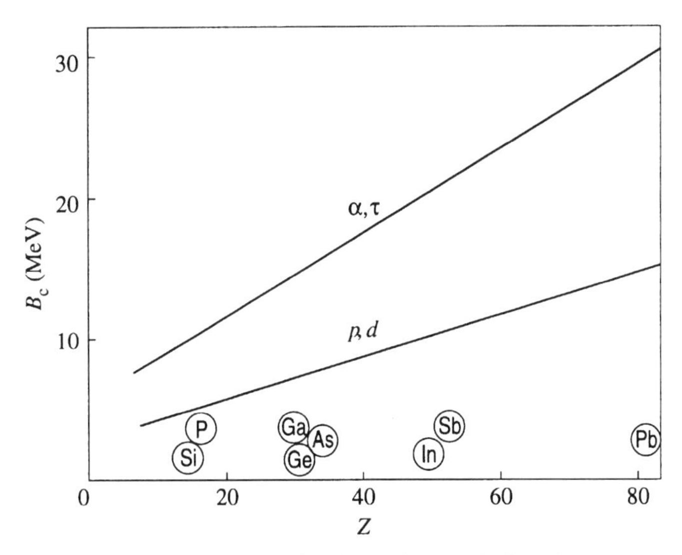

Figure 2.1. Dependence of the Coulomb barrier's height B_C for protons, deuterons, α- and τ-particles on the nuclear charge in the matrix.

nucleus and of the incoming particle, respectively; A_x is the atomic number of the nucleus; e is the charge of the electron. Figure 2.1 shows the dependence of the Coulomb barrier's height B_C for protons and α-particles on the nucleus matrix charge Z. The data are taken from Nemetz and Gofman (1975). As indicated in the figure, B_C produces 5–10 MeV for light nuclei, 10–20 MeV for moderate mass nuclei, and 20–30 MeV for heavy nuclei.

Most nuclear reactions produced by charged particles [(α, n), (α, p), (α, pn), $(\alpha, 2n)$, etc.] are endothermic ($Q < 0$). The nuclear reaction threshold value E_{th} is computed by the equation

$$E_{th} = \frac{M_x + M_a}{M_x} |Q|, \tag{2.2}$$

where Q is the energy of the reaction, M_x is the mass of the nucleus, M_a is the mass of the incoming particle. The value of E_{th} may vary from a few MeV to a few dozens of MeV. Radiation capture reactions (α, γ) and some other reactions, such as those involving deuterons, are exothermic ($Q > 0$), and their threshold energy character depends only on the presence of the Coulomb barrier.

2.2. SIMULATION OF TRANSMUTATION DOPING BY CHARGED PARTICLES

To gain some insight into how impurities are distributed by transmutation doping, we consider the energy transfer from the incoming particles to the atoms in the irradiated material. The kinetic energy of charged particles is expended mainly on ionization and excitation of the target atoms. Therefore along all the way particle flux can be considered relatively constant. The number of nuclear interactions in a thin layer d at depth x from the surface of the target is given by (Nemetz and Gofman 1975)

$$d\nu = F(x)N\sigma(x)\,dx \simeq F_0 N\sigma(x)\,dx, \tag{2.3}$$

where N_0 is the nucleus concentration of the target, F_0 and F are fluxes of the particles near the surface and at depth x, respectively, $\sigma(x)$ is the nuclear interaction cross section. The total number of interactions is equal to

$$\nu = FN\int_0^{R_0} \sigma(x)\,dx = F_0 N \int_{E_{\text{th}}}^{E_0} \frac{\sigma(E)\,dE}{|dE/dx|}, \tag{2.4}$$

where E_0 is the initial energy of particles, R_0 is the depth at which $E = E_{\text{th}}$.

The calculated values of the nuclear reaction thresholds are given by Marples et al. (1967) for the interaction of light charged particles (protons p, deuterons d, ^{3}He nuclei τ, and ^{4}He nuclei α) with different nuclei of the isotopes of the elements encountered in modern semiconductor materials (silicon, germanium, GaAs, InP, GaAlAs, PbTe, CdSe, InAs, PbS, etc.). As a rule the range of possible nuclear reactions considered in the E_{th} calculations is limited to $E_{\text{th}} \sim$ 20–25 MeV. The reason for that is the complexity of the annealing process of radiation defects if the defects are created by particles of higher energies (Smirnov 1977). Figure 2.2 shows an example of how aluminium is formed in silicon under

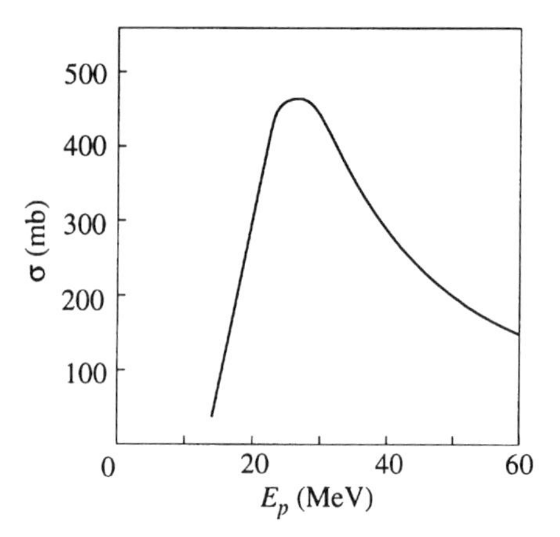

Figure 2.2. Energy dependence of σ, the nuclear reaction cross section of aluminum, under irradiation of ^{28}Si with protons.

irradiation by protons (Mushnikov et al. 1988). As seen in the figure, the maximum of the excitation function (the cross section of forming the aluminium atoms), which is equal to ~ 500 mb, corresponds to the energy ~ 22 MeV.

Figure 2.3 gives the dependences of the reaction cross sections (α, n) and $(\alpha, 2n)$ on the energy of α-particles for $Z_x = 20–40$ (Munzel and Lange 1969). It shows that the maximum cross section of these reactions is $1b$. The maximum occurs where the value of

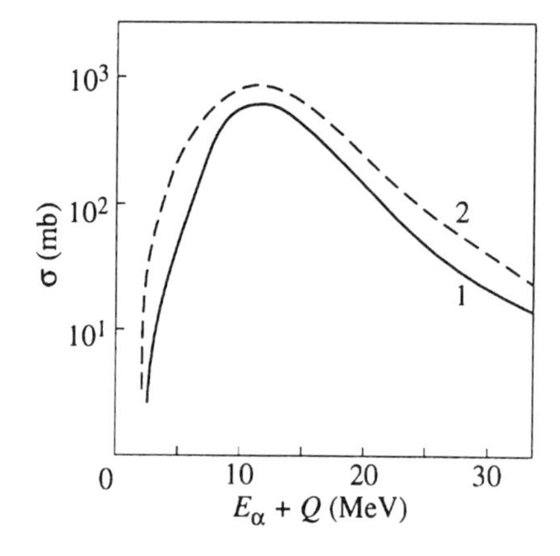

Figure 2.3. Energy dependence of σ, the nuclear reaction cross sections (α, n)—curve 1, and $(\alpha, 2n)$—curve 2—for materials with moderate nuclear charge ($Z = 20–40$) on α-particle energy.

$E + Q$ is equal to 11 MeV. In the absolute energy units the maximum value is observed for the reactions (α, n) at 15 MeV and for $(\alpha, 2n)$ at 25 MeV. At the St. Petersburg Laboratory of Radiation Methods, the simulation of transmutation doping processes for a wide range of semiconductors has been performed (Afonin et al. 1988; Zakharenkov et al. 1990; Zabrodin et al. 1990). The modeling results are given in Tables 2.2 and 2.3. Tables for isotopes of the semiconductor materials contain the values of energy thresholds for different channels of nuclear reactions and the products of those reactions. If the nucleo-product is radioactive, its half-life and the resultant stable nuclei are indicated. Examples of transmutation doping modeling for the irradiation of III-V semiconductor compounds by α- and τ-particles are given in Tables 2.2 and 2.3. The chemical elements, formed in semiconductor materials under proton and α-particle irradiation are generally placed in the periodic table close to the initial elements $(Z' = Z \pm 2)$. By studying these tables, one can establish which elements are due to irradiation by different light ions. Knowing the doping properties of the impurities introduced during transmutation doping, it is possible to predict the effective radiation conditions for doping.

To estimate the total number of doping impurities introduced into a semiconductor, use of the transmutation coefficient (K_{tr}) is recommended (Zatolokin et al. 1976; Dmitriev 1986; Zablotskii et al. 1986). K_{tr} is defined as the ratio of the introduced impurity atoms N_{imp} to the irradiation dose D:

$$K_{tr} = \frac{N_{imp}}{D}. \tag{2.5}$$

The data on the cross sections of the various impurities in semiconductors irradiated by charged particles are scarce. Worth mentioning in this connection is a reference book by Dmitriev (1986) that gives the yields of some radionuclides and the articles by Keller et al. (1973) and Friedlander et al. (1981). But much of the main information about transmutation coefficients must be obtained from experimental work on transmutation doping. Figure 2.4 shows the transmutation coefficients for cadmium selenide (Dmitriev 1986). As is indicated in the figure, the typical values of K_{tr} for the nuclei in the middle of the periodic table are $\sim 10^{-5}$–10^{-6} cm^{-1}.

TABLE 2.2. Main nuclear reactions responsible for the transmutation doping of GaAs, GaAlAs by α- and τ-particles.

Element	Nuclear Reaction	Threshold	Product	$T_{1/2}$	Final Product	E_γ, E_β keV
^{69}Ga	α, n	6.8	^{72}As	26 h	^{72}Ge	$\gamma - 834$
60.7%	$\alpha, 2n$	16.0	^{71}As	65 h	^{71}Ga	$\gamma - 175$
				11 days	^{71}Ga	$\gamma - 175$
	$\alpha, 2p$	12.0	^{71}Ga	Stable	^{71}Ga	—
^{71}Ga	α, n	5.2	^{74}As	18 days	^{74}Ge (68%)	$\gamma - 595$
39.3%					^{74}Se (32%)	$\gamma - 634$
	$\alpha, 2n$	13.6	^{73}As	80 days	^{73}Ge	$\gamma - 53$
	$\alpha, 2p$	13.2	^{73}Ga	4.9 h	^{73}Ge	$\gamma - 297, 326$
^{75}As	α, n	5.5	^{78}Br	6.5 min	^{78}Se	$\gamma - 614$
100%	$\alpha, 2n$	14.2	^{77}Br	58 h	^{77}Se	$\gamma - 818, 239, 520$
	$\alpha, 2p$	11.8	^{77}As	39 h	^{77}Se	$\gamma - 239, 520$
^{27}Al	α, n	3.0	^{30}P	2.5 min	^{30}Si	$\gamma - 2230$ $\beta^+ - 3240$
100%	$\alpha, 2n$	16.0	^{29}P	4 s	^{29}Si	β^+
	$\alpha, 2p$	12.7	^{29}Al	6 min	^{29}Si	$\gamma - 1270$ $\beta^- - 2540$
^{69}Ga	τ, n	−5.7	^{71}As	65 h	^{71}Ge	$\gamma - 175$
				11 days	^{71}Ge	
60.7%	$\tau, 2n$	6.5	^{70}As	52 min	^{71}Ge	$\gamma - 668$ $\gamma - 1040$
	$\tau, 2p$	0.1	^{70}Ga	21 min	^{71}Ge	$\gamma - 176$ $\gamma - 1040$
^{71}Ga	τ, n	−8.0	^{73}As	80 days	^{73}Ge	$\gamma - 53$
39.3%	$\tau, 2n$	3.3	^{72}As	26 h	^{72}Ge	$\gamma - 630$ $\gamma - 834$
	$\tau, 2p$	1.2	^{72}Ga	14 h	^{72}Ge	$\gamma - 630$ $\gamma - 834$
^{75}As	τ, n	−7.4	^{77}Br	57 h	^{77}Se	$\gamma - 239$ $\gamma - 818$
100%	$\tau, 2n$	3.8	^{76}Br	16 h	^{76}Se	$\gamma - 559$ $\gamma - 657$
	$\tau, 2p$	0.4	^{76}As	26 h	^{76}Se	$\gamma - 559$
^{27}Al	τ, n	−7.3	^{29}P	4 s	^{29}Si	β^+
100%	$\tau, 2n$	12.5	^{28}P	0.2 s	^{28}Si	β^+
	$\tau, 2p$	0	^{28}Al	2 min	^{28}Si	$\beta^- - 2870$

Note: Data obtained by radioactive tracer.

TABLE 2.3. Main nuclear reactions responsible for the transmutation doping of InP, InSb by α- and τ-particles.

Element	Nuclear Reaction	Threshold	Product	$T_{1/2}$	Final Product	E_γ, E_β, keV
^{113}In	α, n	8.8	^{116}Sb	16 min	^{116}Sn	$\gamma - 1293, 930$
4.28%	$\alpha, 2n$	16.6	^{115}Sb	32 min	^{115}Sn	$\gamma - 499, 1240$
	$\alpha, 2p$	12.4	^{115}In	Stable	^{115}In	—
	α, n	7.5	^{118m}Sb	5 h	^{118}Sn	$\gamma - 253, 1051, 1229$
^{115}In	$\alpha, 2n$	15.1	^{117}Sb	2.8 h	^{117}Sn	$\alpha - 158, \beta^+$
95.72%	$\alpha, 2p$	13.2	^{117}In	45 min	^{117}Sn	$\gamma - 560, 158$
	α, pn	12.5	^{117m}Sn	14 days	^{117}Sn	$\alpha - 156, 159$
^{31}P	α, n	6.4	^{34m}Cl	32.2 min	^{34}S	$\gamma - 145, 1170$
100%	$\alpha, 2n$	19.4	^{33}Cl	2.5 s	^{33}S	β^+
	$\alpha, 2p$	11.6	^{33}P	25 days	^{33}S	$\beta^- - 250$
^{121}Sb	α, n	8.2	^{124}I	4.2 days	^{124}Te	$\gamma - 602, 722, 1691$
57.25%	$\alpha, 2n$	15.8	^{123}I	13 h	^{123}Te	$\gamma - 159, 530$
	$\alpha, 2p$	13.0	^{123}Sb	Stable	^{123}Sb	—
^{123}Sb	α, n	7.2	^{126}I	13 days	^{126}Te	$\gamma - 386, 860$
42.75%	$\alpha, 2n$	14.5	^{125}I	60 days	^{125}Te	$\gamma - 35$
	$\alpha, 2p$	13.5	^{125}Sb	2.7 yr	^{125}Te	$\gamma - 428, 463, 600, 636$
^{113}In	τ, n	−4.6	^{115}Sb	32 min	^{115}Sn	$\gamma - 499, 1240$
4.28%	$\tau, 2n$	5.8	^{114}Sb	3.3 min	^{114}Sn	$\gamma - 1299$
	$\tau, 2p$	0.4	^{114m}In	49.5 days	^{114}Cd	$\gamma - 192, 558, 724$
^{115}In	τ, n	−6.1	^{117}Sb	2.8 h	^{117}Sn	$\gamma - 158$ $\beta^- - 600$
95.72%	$\tau, 2n$	3.8	^{116}Sb	16 min	^{116}Sn	$\gamma - 1290$
	$\tau, 2p$	0.2	^{116}In	134 s	^{116}Sn	$\gamma - 1097$
	τ, p	−8.7	^{117m}Sn	14 days	^{117}Sn	$\gamma - 156, 159$
^{31}P	τ, n	−3.7	^{33}Cl	2.5 s	^{33}S	β^+
100%	$\tau, 2n$	13.6	^{32}Cl	0.3 s	^{32}S	$\gamma - 2240$
	$\tau, 2p$	−0.2	^{32}P	14 days	^{32}S	$\beta^- - 1710$
^{121}Sb	τ, n	−5.4	^{123}I	13 h	^{123}Te	$\gamma - 159, 530$
57.25%	$\tau, 2n$	4.67	^{122}I	3.5 min	^{122}Te	$\gamma - 564$
	$\tau, 2p$	0.9	^{122}Sb	2.7 days	^{122}Sn	$\gamma - 564, 693$
	τ, p	−7.4	^{123m}Te	120 days	^{123}Sb	$\gamma - 159, 88$
^{123}Sb	τ, n	−6.7	^{125}I	60 days	^{125}Te	$\gamma - 35$
42.75%	$\tau, 2n$	3.2	^{124}I	4.2 days	^{124}Te	$\gamma - 603, 723, 1691$
	$\tau, 2p$	1.3	^{124}Sb	60 days	^{124}Te	$\gamma - 603$
	τ, p	−7.7	^{125m}Te	57 days	^{125}Te	$\gamma - 35, 109$

Note: Data obtained by the radioactive tracer.

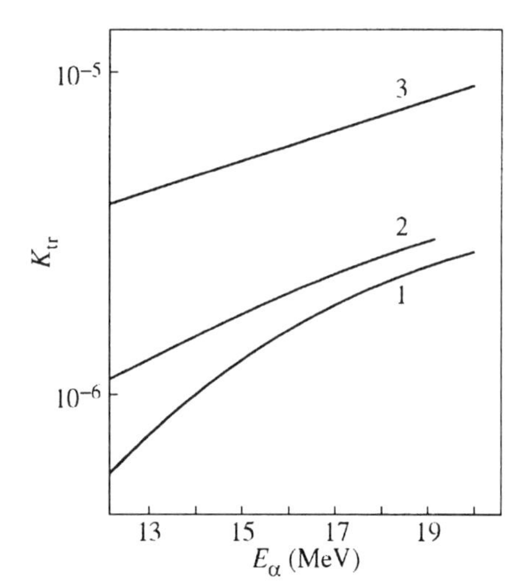

Figure 2.4. Dependence of K_{tr} for the formation of ^{77}Br (1), ^{117m}Sn (2), and ^{113}Sn (3) isotopes in CdSe on α-particle energy.

In transmutation it is difficult to separate the numerous competitive channels of the nuclear reaction leading to the formation of the chosen doping element, especially when the energy of charged particles is high. It is necessary to ensure that the rates of the donor and acceptor impurities introduction are within the required range (Mushnikov et al. 1988). Figure 2.5 shows the dependence of the concentration ratio of acceptor and donor impurities (the compen-

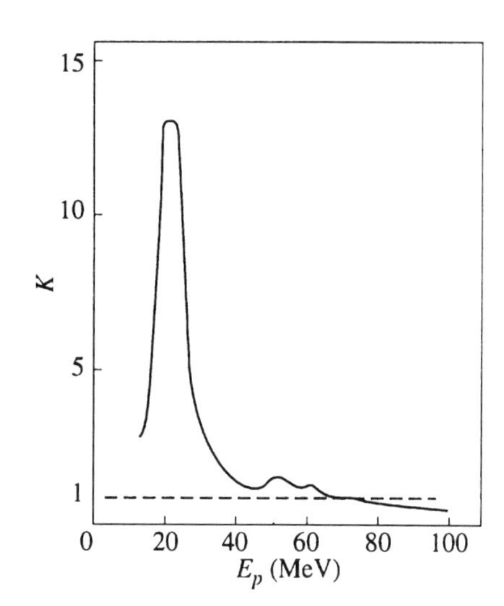

Figure 2.5. Dependence of the concentration ratio of acceptor (aluminum) and donor (magnesium) impurities in silicon on proton energy.

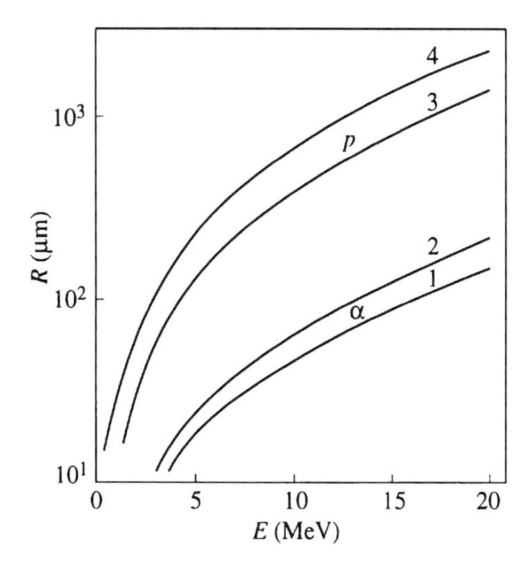

Figure 2.6. Dependence of proton (curves 3, 4) and α-particle (1, 2) ranges R in silicon (2, 4) and GaAs (1, 3) on the ion energy.

sation coefficient) on the energy of protons (Mushnikov et al. 1988) for aluminium and magnesium which are being introduced into silicon. As seen in the figure, the energy of protons $\sim$ 20 MeV is more effective for introducing Al.

When the proton irradiation is used, it is possible to dope a packet of silicon plates whose total thickness is several hundred micrometers. Figure 2.6 shows the energy dependences of the ranges R of the proton- and alpha-particles in silicon and gallium arsenide (Nemetz and Gofman 1975). As the figure shows, the ranges of the protons are about an order of magnitude larger than those of α-particles.

To determine the range of the semiconductor doping more precisely, it is necessary to calculate the depth at which the energy of the charged particle becomes equal to the energy threshold of the nuclear reaction. Figure 2.7 and Table 2.4 show the dependences of the α-particle energy on the penetration depth for different semiconductor compounds (GaAs, InP, GaP, and InAs) (Nemetz and Gofman 1975; Kozlovskii and Abrosimova 1991). The following equation is used in calculating the ranges in complex

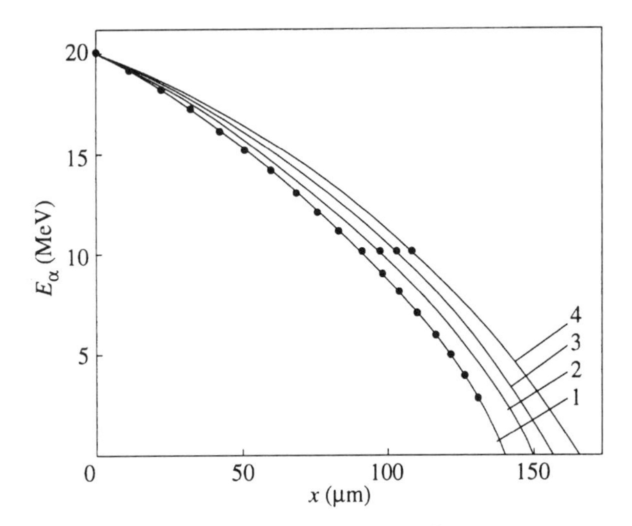

Figure 2.7. Dependence of α-particle energy (for E_0 = 20 MeV) on the depth of penetration into GaAs (1), InAs (2), GaP (3), and InP (4).

semiconductor compounds:

$$\frac{1}{R} = \frac{w_1}{R_1} + \frac{w_2}{R_2} + \frac{w_3}{R_3} + \cdots, \tag{2.6}$$

where R is the range in a chemical compound, $R_1, R_2, \ldots, R_n$ are ranges of paths in different elements that the compound consists of, $w_1, w_2, \ldots, w_n$ are the relative shares of the elements in the compound (by mass). As indicated in the figure, the variations of penetration depth of α-particles with minimal energy is 20 MeV in different compounds $A^{III}B^{V}$ are around 10 percent (GaAs−141 μm, InAs−150 μm, Gap−158 μm, InP−165 μm) (Didik 1993a).

The threshold values of nuclear reactions can be used to predict that the characteristic depths of transmutation doping of semiconductor compounds, when irradiated by α-particles, will reach several microns. It is also possible to form different impurity profiles: either one with a highly uniform impurity distribution or one with an extremum, smoothly falling, and so on.

TABLE 2.4. Dependences of the energies of α-particles in the major III-V compounds on their penetration depth (in microns) into a semiconductor for $E_0 = 20$ MeV.

Energy, MeV	GaAs	InP	GaP	InAs
20	0	0	0	0
19	11.1	13.0	12.4	11.8
18	21.8	25.5	24.2	23.2
17	32.1	37.6	36.0	34.2
16	42	49.2	47.1	44.6
15	51.5	60.4	57.7	54.8
14	60.6	70.9	67.9	64.5
13	69.3	81.2	77.7	73.8
12	77.6	90.7	86.9	82.5
11	85.5	100.0	95.7	91.0
10	92.9	109.0	104.0	99.0
9	99.9	117.0	112.0	106.0
8	106.4	125.0	119.0	113.0
7	112.5	131.0	126.0	120.0
6	118.1	138.0	132.0	125.0
5	123.2	144.0	138.0	131.0
4	127.8	150.0	143.0	136.0
3	131.9	154.0	148.0	140.0
2	135.2	158.0	152.0	144.0
1	138.2	162.0	155.0	147.0
0	141.0	165.0	158.0	150.0

2.3. EXPERIMENTAL INVESTIGATION OF TRANSMUTATION DOPING BY CHARGED PARTICLES

The basic experiments are performed with beams of light nuclei of hydrogen and gelium: protons, deutrons, nuclei of gelium-3 (τ-particles), and gelium-4 (α-particles). The choice of nuclei is determined by two factors.

First, the thresholds of nuclei reactions, whose products are the nuclei of the requisite doping impurities, are $\sim$ 5–15 MeV for light nuclei and more than 30–40 MeV for medium nuclei (by mass and charge). The task of obtaining larger fluxes of light charged particles (up to 10^{18} cm^{-2}) accelerated to $\sim$ 10 MeV is much easier than that of obtaining similar fluxes of charged particles, medium by mass, accelerated to > 30 MeV. Second, unlike the medium and heavy charged particles, light particles of 10–20 MeV create mainly point or binary defects which can be easily eliminated by subsequent annealing (Smirnov 1977; Lang 1977; Kozlovskii and Zakharenkov 1995).

2.3.1. Silicon

The first experiments on silicon were made by Trey and Oberhauser (1957) and Dolgolenko and Shakhovtsev (1970). Both studied the introduction of donor impurities (phosphorus and sulfur) into silicon by irradiating it with α-particles and forming a p-n junction in p-Si ($\rho \sim 30\ \Omega \cdot \text{cm}$). Estimates show that it is necessary to have the α-particle dose of $\sim 7 \cdot 10^{17}$ cm^{-2} in order to obtain a donor concentration of about 10^{15} cm^{-3}. For shallow p-n junctions ($\sim 7\ \mu$m), α-particles with energy of 5.3 MeV (Trey and Oberhauser 1957) were used. For deep junctions (200–300 μm), the α-particles must have the energy of 27.2 MeV (Dolgolenko and Shakhovtsev 1970).

Without the forced cooling, the temperature of the specimens being irradiated rose to 900°C. According to Dolgolenko and Shakhovtsev (1970), this enabled annealing the radiation defects and removing the implanted helium from silicon. As a result of those experiments, it became possible to form a p-n junction. It was noticed that the greater the irradiation dose, the deeper is the doped region up to the fixed level of $\sim 10^{15}$ cm^{-3}.

The annealing that was performed (950°C, 2 h) proved insufficient for a complete removal of the helium-vacancy complex, though most helium was removed from the specimen. Some publications were devoted to the experimental investigation of transmutation doping of silicon by the acceptor impurity (aluminium) by charged particles (Gaidar et al. 1986; Varnina et al. 1981). Gaidar et al. (1986) irradiated n-silicon samples with the electron concentration $(2\text{–}3) \cdot 10^{12}$ cm^{-3} by protons with energy of 25 MeV, dose $2 \cdot 10^{16}$ cm^{-2}, and then annealed the samples at 850°C for 1.5 hours. The main nuclear reactions, resulting in the formation of aluminium,

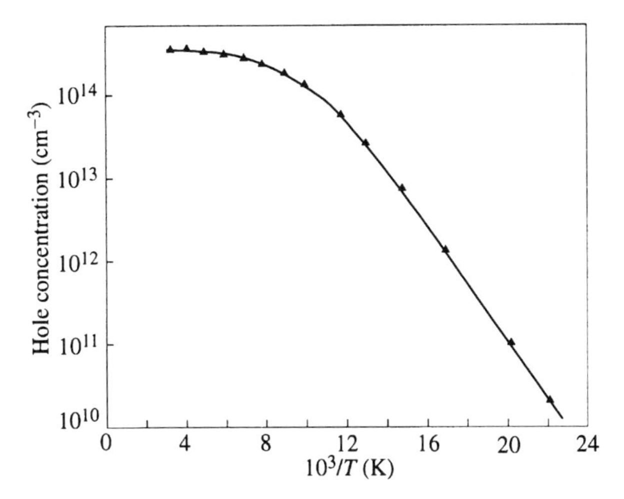

Figure 2.8. Temperature dependence of hole concentration in silicon transmutationally doped with aluminum.

are the following:

	E_{th}	MeV	
$^{29,30}Si(p, \alpha)^{26,27}Al$	5.0	2.5	
$^{28}Si(p, pn)^{27}Si \xrightarrow[T]{\beta^+} {}^{27}Al$	15.7		(2.7)
$^{28}Si(p, 2p)^{27}Al$	12.0		

Figure 2.8 shows the temperature dependences of the hole concentration, obtained from measuring the temperature dependences of resistivity (Gaidar et al. 1986). As is seen in Figure 2.8, the mentioned conditions of the transmutation doping of silicon lead to the formation of the acceptor impurity (aluminium) with an effective hole concentration of $\sim 4 \cdot 10^{14}$ cm^{-3}.

Varnina et al. (1988) primarily devote their investigation to defect formation in silicon transmutationally doped by aluminium, after irradiation by protons of 40–50 MeV. These authors show that the rates of introducing disordered regions and filling vacancies during proton and neutron irradiation are similar in magnitude. The rates of introducing the A centers during the proton irradiation are two or three orders of magnitude higher. This research shows that using the pulse annealing of one-second duration helps eliminate the radiation defects.

The work performed in the St. Petersburg Laboratory of Radiation Methods should be mentioned as well (Gornushkina et al. 1991). It was the first study of the transmutation impurity depth distribution. The donor impurities (sulphur, phosphorus) were formed in silicon irradiated by α-particles with energies of 12, 16, and 20 MeV. The radioactive impurity distribution was determined by removing thin Si layers and measuring the remaining β activity of the sample. Figure 2.9 shows the distribution of phosphorus ^{32}P, formed by the nuclear reaction $^{29}Si(p, \alpha)^{32}P \rightarrow {}^{32}S$, $E_{th} = 2.8$ MeV after the irradiation by α-particles with energies of 12, 16, and 20 MeV (Gornushkina et al. 1991). Irradiation at oblique angles was used (the slope of the beam to the surface of the specimen being 30, 18, and 6 degrees) in order to control the thickness of the doped layer. As can be seen in Figure 2.9, doped layers with thicknesses from several microns to hundreds of microns are formed by the irradiation. The annealing studies within a wide temperature range showed that at more than 1100°C and after 50 hours, a

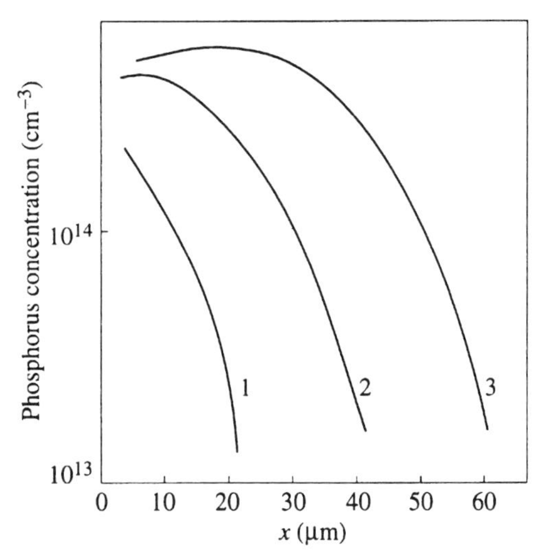

Figure 2.9. Distribution of the phosphorus isotope ^{32}P in silicon after irradiation with α-particles of energies 12 MeV (1), 16 MeV (2), and 20 MeV (3). The beam's angle of incidence on the sample's surface is 18°.

thermal diffusion starts taking place, with the diffusion coefficient in the transmutationally doped silicon exceeding the corresponding coefficient in an unirradiated semiconductor four to five times.

2.3.2. Semiconductor Compounds $A^{III}B^{V}$

The effective introduction of donor impurities was mentioned in early publications on transmutation doping of $A^{III}B^{V}$ compound ingots irradiated with neutrons (Mirianashvili et al. 1976; Vodopyanov and Kurdiani 1966; Mirianashvili and Nanobashvili 1971; Vavilov et al. 1976; Prussin and Cleland 1978).

In the 1980s the development of the technology of power and galvanomagnetic devices gave rise to the interest in creating nonuniform impurity profiles and p-n structures.

At the Laboratory of Radiation Methods a number of experiments were carried out in this field. The results of these experiments constitute a separate section of this chapter. The publication by Afonin et al. (1988) contains an analysis of impurity composition in GaAs irradiated with protons and α-particles with energy of 15–20 MeV.

It was shown that the most efficient nuclear doping of GaAs is that by the donor impurities (Ge, Se) irradiated with α-particles. Zabrodin et al. (1990) reported on the high efficiency of α-irradiation when InP is transmutationally doped by the donor impurities (S, Sn). Both experiments were performed on the MGC-20 cyclotron. The samples were irradiated by protons of 18 MeV and α-particles of 20 MeV, with the dose of $2 \cdot 10^{16}$ cm^{-2} at room temperature. The carrier concentration after the irradiation was determined by capacitance-voltage method using a Schottky barrier probe. The amount of impurities formed from the radioactive products of nuclear reactions was determined by measuring the γ activity of the samples.

Table 2.5 lists the experimental values of transmutation coefficients for main nuclear reactions in $A^{III}B^{V}$ compounds. Figure 2.10 shows the dependence of these coefficients on the α-particle energy (Zakharenkov et al. 1990). As shown in Figure 2.10 and Table 2.5, the values K_{tr} mostly lie within the range of 10^{-4} to 10^{-6} cm^{-1}. One can expect that the K_{tr} for analogous nuclear reactions are

TABLE 2.5. Experimental values of the transmutation coefficients of the main nuclear reactions.

Particle	Nuclear Reaction	Dopant	Threshold, MeV	$K_{tr} \cdot 10^5$
	$^{69}Ga(\alpha, n)^{72}As$	^{72}Ge	6.8	5
	$^{71}Ga(\alpha, n)^{74}As$	^{74}Ge, ^{74}Se	5.2	4
	$^{75}As(\alpha, 2p)^{77}As$	^{77}Se	11.8	4
	$^{75}As(\alpha, 2n)^{77}Br$	^{77}Se	14.2	4
	$^{31}P(\alpha, 2p)^{33}P$	^{33}S	11.6	3
α	$^{115}In(\alpha, n)^{118m}Sb$	^{118}Sn	7.5	0.7
$^4He^{2+}$	$^{115}In(\alpha, pn)^{117m}Sn$	^{117}Sn	12.5	0.4
	$^{121}Sb(\alpha, n)^{124}I$	^{124}Te	8.2	1
	$^{121}Sb(\alpha, 2n)^{123}I$	^{123}Te	15.8	2
	$^{123}Sb(\alpha, n)^{126}I$	^{126}Te, ^{126}Xe	7.2	1
	$^{123}Sb(\alpha, 2n)^{125}I$	^{125}Te	14.5	1
	$^{69}Ga(\tau, 2p)^{70}Ga$	^{70}Ge	0.1	0.3
	$^{69}Ga(\tau, 2n)^{70}As$	^{70}Ge	6.5	1
	$^{71}Ga(\tau, 2p)^{72}Ga$	^{72}Ge	1.2	0.5
	$^{71}Ga(\tau, 2n)^{72}As$	^{72}Ge	3.3	2
	$^{71}Ga(\tau, n)^{73}As$	^{73}Ge	−8.0	0.2
τ	$^{75}As(\tau, 2n)^{76}Br$	^{76}Se	3.8	3
$^3He^{2+}$	$^{75}As(\tau, n)^{77}Br$	^{77}Se	−7.4	0.1
	$^{75}As(\tau, 2p)^{74}As$	^{76}Se	0.4	0.4
	$^{31}P(\tau, 2p)^{32}P$	^{32}S	−0.2	0.3
	$^{115}In(\tau, p)^{117m}Sn$	^{117}Sn, ^{122}Te	−8.7	0.2
	$^{121}Sb(\tau, 2p)^{122}Sb$	^{125}Te	0.9	0.3
	$^{123}Sb(\tau, n)^{125}I$	^{124}Te	−6.7	0.1
	$^{123}Sb(\tau, 2n)^{124}I$	^{125}Te	3.2	0.7
	$^{123}Sb(\tau, p)^{125m}Te$	^{124}Te	−7.7	0.1
	$^{123}Sb(\tau, 2p)^{125}Sb$		1.3	0.2

similar to those listed in Tables 2.2, 2.3, and 2.5. As a result of the irradiation by protons, both acceptor (Zn) and donor (Se, Ge) impurities are present in GaAs. Their transmutation coefficients are quite small ($K_{tr} \sim 10^{-6}$). The difference between the impurity concentrations is also very small.

Unlike the proton irradiation the irradiation of GaAs by α-particles results in the formation of only donor impurities (Ge, Se). As this takes place, the number of such nuclear reactions is about 15,

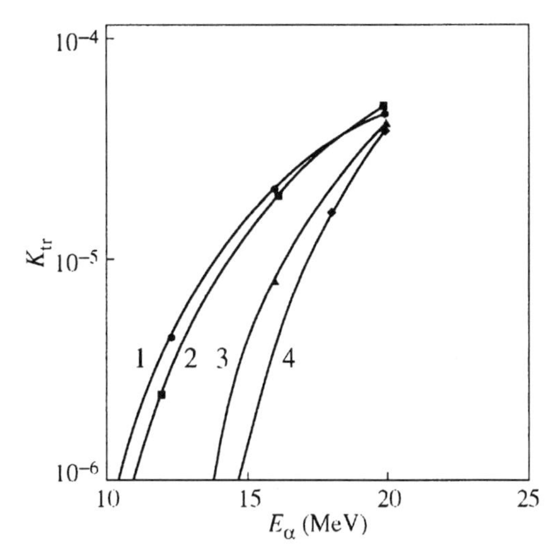

Figure 2.10. Dependence of the transmutation coefficient K_{tr}, representing the formation of ^{74}As (1), ^{72}As (2), ^{73}As (3), and ^{77}Br (4) isotopes, on α-particle energy.

while the energy of α-particles is equal to 20 MeV. According to data obtained by Afonin et al. (1988) and Zakharenkov et al. (1990), approximately $\sim 4 \cdot 10^{12}$ cm^{-2} atoms of the donor impurity are introduced into the semiconductor when GaAs is irradiated with α-particles with energy of 20 MeV and dose $2 \cdot 10^{16}$ cm^{-2}. The total K_{tr} value of the donor impurities introducing under these conditions is $2 \cdot 10^{-4}$ cm^{-1}. Since the penetration depth under such conditions is on the order of ~ 100 μm, the concentration of the doping impurities irradiated with the $2 \cdot 10^{-16}$ cm^{-2} dose is close to $4 \cdot 10^{14}$ cm^{-3}.

Measurements of the carrier concentration in these samples after the annealing at a temperature of 750°C for 30 minutes indicated a surface concentration of $2.5 \cdot 10^{14}$ cm^{-3}, which agrees well with the value determined by the radioactivity of the introduced impurity. The difference between the donor impurity concentration (Ge, Se) and the carrier concentration can be explained, first, by the radiation-stimulated diffusion of the impurity that occurs during the annealing of the radiation defects (Dzhafarov 1991) and, second, by the ambiguous localization of Ge atoms in the gallium sublattice, which has been noted in publications on neutron doping of GaAs (Garrido et al. 1985; Kolin et al. 1984; Kolin et al. 1987; Bykovskii et al. 1989).

Zakharenkov et al. (1990) and Zabrodin et al. (1990) in their studies of InP transmutation doping come to the conclusion that it is quite effective to use radiation for doping InP with donor impurities. The values of K_{tr} obtained experimentally were $2 \cdot 10^{-4}$ for $E_\alpha = 20$ MeV; for protons of 18 MeV, K_{tr} was less than 10^{-5}. With equal doses of radiation ($2 \cdot 10^{16}$ cm^{-2}) the carrier concentration was $8 \cdot 10^{12}$ cm^{-3} for protons and $4 \cdot 10^{14}$ cm^{-3} for α-particles (Zabrodin et al. 1990). A decrease of the bombarding particle energy showed a decrease of the K_{tr} value as well. The case where GaAs is irradiated with α-particles of different energies appears in Figure 2.10, along with the measured dependence of the transmutation coefficient for the formation of isotopes ^{72}Ge, ^{74}Ge, ^{74}Se, ^{73}Ge, ^{77}Se (Zakharenkov et al. 1990). As is seen in the figure, when $E < B_C(\approx 15$ MeV), the nuclear reactions still occur, and the reaction cross sections tend to zero when E_α is close to the nuclear reaction threshold.

Profiles of the doping impurity distribution are of a special interest. Some publications have been devoted to the investigation of the concentration profiles of the impurities decaying into nuclei of the donor atoms (Didik et al. 1989; Didik et al. 1990; Didik et al. 1993a; Didik et al. 1993b). They investigated the profiles of selenium, germanium, tin, sulphur, and other impurities in GaAs, InAs, GaP, and InP. Figures 2.11 and 2.12 show the cases of InP and GaAs. Figure 2.11 shows that the depth of the layer of InP, doped with sulphur and irradiated with α-particles whose energy is 20 MeV (curve 1), is close to ~ 70 μm, which agrees well with the threshold of the corresponding nuclear reaction (11.6 MeV) and with the energy of the α-particles at that depth (Didik et al. 1990).

It has been established that the energy of the particles can affect not only the doped layer depth but also the shape of the doping profile (Didik et al. 1989). Figure 2.12 shows the distribution of isotope ^{72}As in gallium arsenide depending on the energy of α-particles. As seen in the figure, for the energies of 12 and 16 MeV the impurity distributions are described by falling curves, and for $E = 20$ MeV the curve has a maximum (Didik et al. 1993b). To explain these curves, we refer to Figure 2.3. The maximum (α, n)

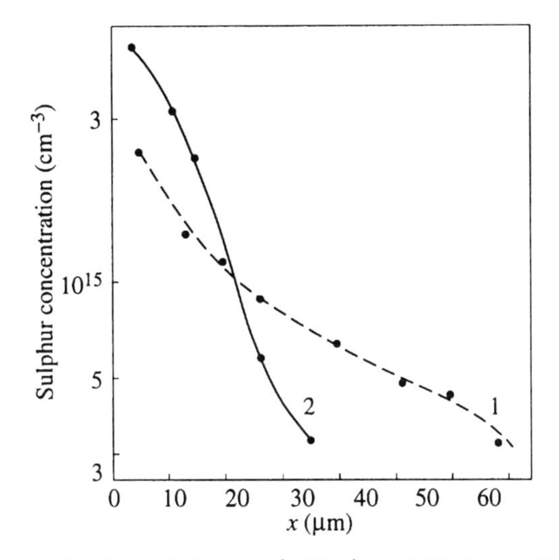

Figure 2.11. Distribution of donors (sulfur) in InP, formed by transmutation doping with α-particles of 20 MeV. The angles of incidence of the beam on the sample are 90° (1) and 30° (2).

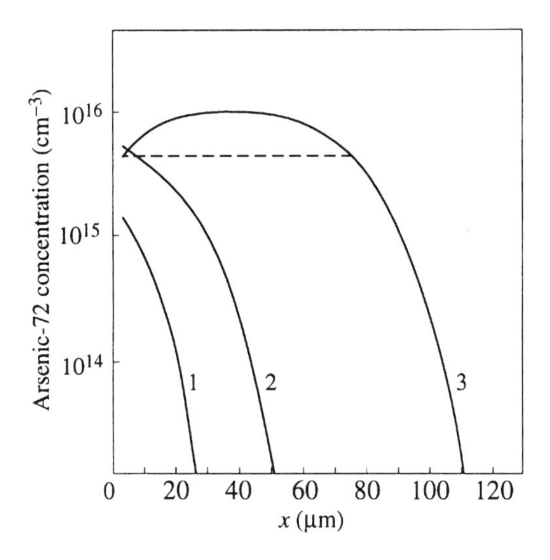

Figure 2.12. Effect of α-particle energy on the distribution of the $^{72}As \Rightarrow {}^{72}Ge$ isotope formed in GaAs by the reaction of $^{69}Ga(\alpha, n)^{72}As$ with a reaction threshold of 6.8 MeV. E (MeV): 12, 16, and 20 (curves 1, 2, and 3, respectively).

and $(\alpha, 2n)$ reaction cross sections occur when the energies of α-particles are greater than the threshold of the nuclear reaction E_{th} by 11 MeV. For the isotope ^{72}As the nuclear reaction ^{69}Ga (α, n) ^{72}As has the E_{th} value equal to 6.8 MeV. Hence the maximum cross section of the reaction corresponds to $E = 18$ MeV. Alpha-particles with the initial energy of 20 MeV are stopped at the energy of 18 MeV at approximately 30-μm depth (see Figure 2.7). This explains the absence of maxima at $E_0 = 16$ and 12 MeV (see Figure 2.12, curves 1 and 2).

Annealing conditions greatly affect the properties of radiation defects and the electric activation of impurities. In the work on neutron doping it has been noted that the main radiation defects are annealed at ~ 700°C and that complete elimination of defects can be observed at ~ 1000°–1100°C (Mirianashvili and Nanobashvili 1971; Prussin and Cleland 1978; Kolin et al. 1988). For doping the $A^{III}B^{V}$ compounds, the same annealing regimes have generally been used but without a careful investigation of the

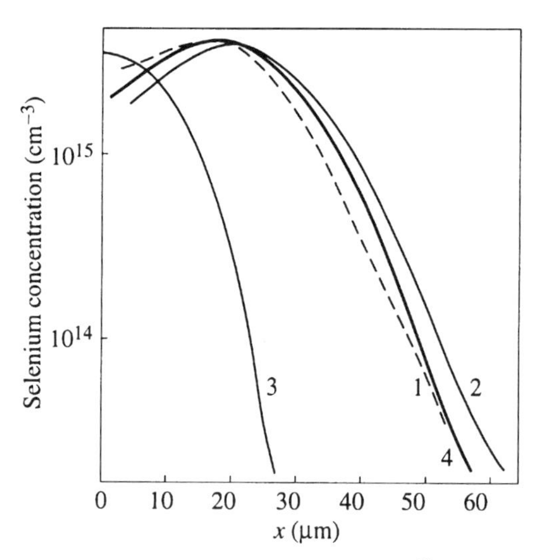

Figure 2.13. Distribution of the donor impurity ^{74}Se in gallium arsenide before (1) and after annealing at 1160°C for 2 h (2) and 20 h (3, 4). *Annealing conditions*: Vacuum (curves 2, 3), doped with a small amount of arsenic amounting to 20 mg (4).

presence of defects in the material. Some works have measured separately the concentrations of the introduced impurities and those of the electrically active impurities in different regimes of annealing (Afonin et al. 1988; Zakharenkov et al. 1990; Zabrodin et al. 1990). These measurements have led to recommendations on regimes suitable for GaAs and InP.

High-temperature annealing of GaAs, transmutationally doped with selenium and germanium, has also been investigated (Didik et al. 1993b). These experiments have established the maximum annealing temperature at which the distortion of the doping impurity profiles due to thermal diffusion is minimal with no disruption of stoichiometry due to evaporation of the components. The experiments were conducted in evacuated ampoules both with and without the addition of arsenic, at temperatures up to 1160°C and durations up to 17 hours. The results (Didik et al. 1993b) are reviewed in Figure 2.13. It is clear that when the duration of annealing is 2 hours and the maximum temperature is 1160°C, the distortions of the impurity profile are minor. If the annealing lasts longer, it is necessary to add an arsenic charge.

2.3.3. Other Materials

In principle, the doping of germanium by α-particles and deutrons was described as far back as in the 1940s (K. Lark-Horowitz et al. 1948). Under irradiation germanium can be doped with such impurities as Ga, Se, As, and Br. However, mainly these early experiments introduced arsenic by the reactions ^{74}Ge(d, p) ^{75}Ge $\rightarrow$ ^{75}As, ^{74}Ge(d, n) ^{75}As and ^{72}Ge (α, n) ^{75}Se $\rightarrow$ ^{75}As, ^{73}Ge($\alpha, 2n$) ^{75}Se $\rightarrow$ ^{75}As. Such irradiation facilitates forming the material with n-type conductivity. The extended radioactivity of the irradiated samples is an essential obstacle to the nuclear doping of germanium.

Some works have suggested the possibility of transmutation doping of silicon carbide with charged particles, both with the acceptor (aluminium) and donor impurities (phosphorus, sulphur) (e.g., see Mokhov et al. 1992). The formation of a doped layer, several microns deep, was obtained even when temperatures were reduced abruptly: from 2500°–3000°C to 1000°–1200°C.

The potential for transmutation doping was studied for a number of isotopes in copper pyrites $CuInSe_2$, a promising solar cell material (Didik et al. 1994b). The irradiation was done with protons, deutrons, and α-particles. By varying particle type and energy, it was possible to control the introduced impurity and the depth of doping.

Recently the transmutation doping of other materials, such as ferroelectric $PbZrTIO_3$ films and high-temperature superconducting materials, has produced promising results in our laboratory radiation studies. Doping by high-temperature superconducting ceramics $YBa_2Cu_3O_7$ has also been studied after irradiation of α-particles with energy of 20 MeV. Figure 2.14 shows the distribution depth of isotopes ^{140}La, ^{92m}Nb, ^{141}Ce, ^{67}Ga, ^{66}Ga, and ^{67}Cu after the material had been irradiated with a dose of 10^{16} cm^{-2} (Didik et al. 1990). As the figure shows, the formation of impurity atoms (lanthanum, cesium, zinc, etc.) takes place up to $\sim$ 100 μm. The profiles of the impurity distribution are similar in shape, except for curve 3, which reflects the distribution of isotope ^{92m}Nb. Isotope ^{92m}Nb is the

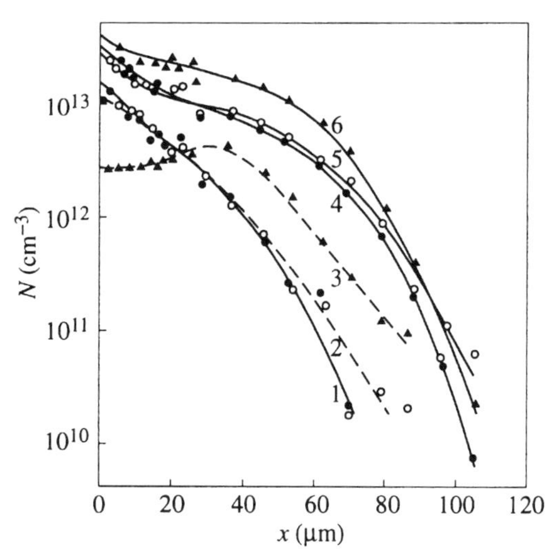

Figure 2.14. Distribution profiles of the isotopes ^{140}La (1), ^{141}Ce (2), ^{92m}Nb (3), ^{67}Ga (4), ^{67}Cu (5), and ^{66}Ga (6) after irradiation of a high-temperature superconducting ceramic by α-particles with 20 MeV energy.

product of the reaction $^{89}Y(\alpha, n)\ ^{92m}Nb$ which has a low energy threshold $E_{th} \sim 7$ MeV and therefore (as has been shown before) is characterized by a concentration maximum at the impurity distribution profile.

The diffusion annealing of polycrystal specimens doped by the transmutation method are of special interest. In the diffusion coefficients of impurities introduced into ceramics one can expect a considerable diversion from the surface source between the transmutation doping method with charged particles and traditional thermal diffusion. The difference between the diffusion coefficients demonstrates the different diffusion mechanisms. When diffusion is thermal, the impurity atoms are moved mainly along the ceramic's grains and pores, whereas in transmutation doping the limiting process is that of impurity diffusion within the ceramic's grains. Thus separate estimations of the impurity diffusion parameters can be done for the grain borders, the ceramic's pores, and the inside of the grains.

A careful study of the concentration profiles of the impurity distribution $C(x)$ enables one to solve the inverse problem of determining energy dependence of the nuclear reaction cross section $\sigma(x)$.

From Eq. (2.3) we have

$$C(x) = NF\sigma(x)\lambda^{-1}(1 - e^{-\lambda t}), \tag{2.8}$$

where F is a flux of particles, λ is a decay constant of radionuclide, and t is the duration of irradiation. This method has allowed one to obtain experimentally the cross sections of the nuclear reactions for a large group of isotopes (Didik et al. 1994a). This application is especially important for determining $\sigma(x)$ when a thin target method, quite standard in nuclear physics, cannot be used.

2.4. POTENTIAL OF THE METHOD IN THE TECHNOLOGY OF SEMICONDUCTOR DEVICES

An analysis of the applicability of the three radiational methods of doping listed in Table 2.1 (which excludes transmutation doping) is given by Kozlovskii and Lomasov (1985) and Dzhafarov (1991).

They point out that the most effective methods for manufacturing microwave and optoelectronic devices are radiation-stimulated diffusion and doping by radiation defects during irradiation with protons at 100 keV. This choice was determined by the submicron depth of the radiation effect in the above methods of doping.

Transmutation doping with charged particles is the most efficient method for forming deep (several hundreds of microns) lightly doped layers. Such layers are used for manufacturing power devices. A method of manufacturing high-voltage GaAs diodes ($U_{op} \sim 1$ kV) has been described by Kozlovskii (1991). For such a device it is necessary to form two lightly doped layers with a doping level less than $5 \cdot 10^{15}$ cm^{-3} and more than 150 μm thick. For this purpose we can use either α- or τ-particle irradiation (Kozlovskii and Abrosimova 1991). Both the energy and the dose of the irradiation are determined by certain characteristics of the diode. Kozlovskii and Abrosimova also provide a description of gallium arsenide doping with selenium and germanium by irradiating GaAs with α-particles in order to create n-n$^+$ junctions of galvanomagnetic devices at the depth of ~ 10 μm with the mean carrier concentration in the layer of $\sim 10^{15}$ cm^{-3}. To obtain thin doped layers, we recommend use of oblique beams rather than decreasing the energy of bombarding particles, since any decrease also reduces the transmutation coefficient.

To create deep p-channels in silicon by irradiating it by protons with energies of 20–25 MeV, we recommend use of transmutation doping with charged particles (Gaidar et al. 1986). In this case the screens used during the irradiation allow completely vertical p-channel borders to form in the local parts of the plate. The placement of aluminium on the surface is determined by two factors: by the angular divergence of the incident proton beam and by the diffusion of aluminium under the mask during annealing of the radiation defects. Since the divergence of the beam is not more than several degrees and the elimination of defects can be performed by pulsed annealing, transmutation doping can be used for the formation of deep channels in the local semiconductor regions.

Abvil et al. (1994) have proposed using transmutation doping to form a phosphorus impurity layer in silicon far below the surface. Using the resonance reaction (i.e., a reaction with a pronounced dependence $\sigma(E)$), of the radiational capture ^{30}Si(p, γ)^{31}P allows doped layers to form that are a fraction of a micron thick at the distances of 10–20 μm from the surface. In order to form a similar layer by implanting phosphorus ions, the initial energy of the ions should cxceed 15 MeV. Abril et al. also note that transmutation doping makes it possible to decrease the concentration of the introduced radiation defects and to form much thinner impurity layers.

2.5. CONCLUSION

The main feature of transmutation doping by charged particles, compared to other methods of transmutation doping, is its ability to create nonuniform impurity profiles forming, for example, a n-n$^+$ or a p-n structure. The depth of the doped layer is determined by the energy of the charged particles, by their stopping losses in the semiconductor, and by the value of the nuclear reaction threshold. Second, unlike the methods of neutron- and photo-nuclear transmutation doping, this technique makes it possible to create a wide spectrum of impurities with both conductivity types. Third, only this method allows one to selectively introduce the impurity into certain surface regions of the semiconductor by using screens, masks, or special beam focusing.

The concentration of the impurities introduced by charged particles is determined by the values of the cross sections of the main nuclear reactions. The transmutation coefficients are equal to 10^{-3}–10^{-4} cm^{-1}. Experiments have shown that this method is mainly applicable for the creation of lightly doped regions of the semiconductor. The shape of the impurity distribution profile is determined by the energy dependence of the cross section of the relevant nuclear reaction. By regulating the conditions of the irradiation, it is possible to form profiles of different configuration: with an increased uniformity of impurity distribution, with extrema, gradually descending, and so on. The total length of the doped layer can vary from a fraction of a micron to hundreds of microns.

The authors of this chapter thank R. Sh. Malkovich and V. A. Didik for useful discussions of problems of transmutation doping by charged particles and B. A. Shustrov for the constructive criticism.

REFERENCES

Abril, I., Garsia-Molina, R., Erokhin, K. M., and Kalashnikov, N. P. (1994). In *Proc. XXIV All-Union Council on Interaction Physics of Charged Particles with Crystals*, p. 77. MSU, Moscow (in Russian).

Afonin, O. F., Viktorov, B. V., Zabrodin, B. V., et al. (1988). *Sov. Phys. Semicond.* 22, 35–38.

Bykovski, V. A., Girii, V. A., Korshunov, F. P., and Utenko, V. I. (1989). *Sov. Phys. Semicond.* 23, 48–50.

Didik, V. A., Kozlovskii, V. V., Malkovich, R. Sh., Skoryatina, E. A., and Shustrov, B. A. (1989). *Sov. Tech. Phys. Lett.* 15, 462–463.

Didik, V. A., Kozlovskii, V. V., Malkovich, R. Sh., Skoryatina, E. A., and Shustrov, B. A. (1990). In *Proc. Int. Conf. on Radiation Materials Science* 6, Alushta, Crimea, pp. 50–53 (in Russian).

Didik, V. A., Kozlovskii, V. V., and Mokhov, V. N. (1992). *Sov. Tech. Phys. Lett.* 18, 191–192.

Didik, V. A., Kozlovskii, V. V., Malkovich, R. Sh., and Skoryatina, E. A. (1993a). *Semicond.* 27, 148–149.

Didik, V. A., Kozlovskii, V. V., Malkovich, R. Sh., and Skoryatina, E. A. (1993b). *Semicond.* 27, 189–190.

Didik, V. A., Kozlovskii, V. V., Malkovich, R. Sh., and Skoryatina, E. A. (1994a). *At. Energ.* 77, 81–83 (in Russian).

Didik, V. A., Kozlovskii, V. V., Malkovich, R. Sh., and Skoryatina, E. A. (1994b). *Sov. Tech. Phys. Lett.* 20, 672–673.

Dmitriev, P. P. (1986). In *Radionuclide Yield of Reactions with Protons, Deuterons, Alpha-Particles and Helium*-3, Moscow, Energoatomizdat, pp. 141–148 (in Russian).

Dolgolenko, A. P. and Shakhovtsev, V. I. (1970). In *Radiation Physics of Nonmetallic Crystals*. Minsk, Nauka i Technika, pp. 191–194 (in Russian).

Dzhafarov, T. D. (1991). In *Radiation-Stimulated Diffusion in Semiconductors*, pp. 134–137. Energoatomizdat, Moscow (in Russian).

Friedlander, G., Kennedy, J. W., Macias, E. S., and Miller, J. W. (1981). In *Nuclear and Radiochemistry*, pp. 294–297. Wiley, New York.

Gaidar, G. P., Dmitrenko, N. N., Dubar, I. V., et al. (1986). *Sov. Phys. Semicond.* 20, 607–608.

Garrido, J., Castano, J. L., Piqueras, J., and Alcober, V. (1985). *J. Appl. Phys.* 57, 2186–2190.

Goldin, L. L. (1983). In *Physics of Accelerators*, pp. 144–145. Nauka, Moscow (in Russian).

Gornuschkina, E. D., Didik, V. A., Kozlovskii, V. V., and Malkovich, R. Sh. (1991). *Sov. Phys. Semicond.* 25, 1232–1233.

Keller, K. A., Lange, J., Muncel, H., and Pfennig, G. (1973). In *Ladolt-Borstein Numerical Data and Functional Relationships in Science and Technology*, Hellwege, K., and Schopper, H., eds., pp. 174–178. Springer-Verlag, Berlin.

Kolin, N. G., Kulikova, L. V., Osvenski, B. V., et al. (1984). *Sov. Phys. Semicond.* 18, 1364–1366.

Kolin, N. G., Kol'chenko, T. I., and Lomako, V. M. (1987). *Sov. Phys. Semicond.* 21, 197–198.

Kolin, N. G., Kulikova, L. V., and Osvenski, B. V. (1988). *Sov. Phys. Semicond.* 22, 646–648.

Komar, E. G. (1975). In *Fundamentals of Accelerated Technology*, pp. 368–370. Moscow (in Russian).

Kozlovskii, V. V. and Lomasov, V. N. (1985). In *Obz. Electron. Tekhn.* Ser. 7, 9, 3–56 (in Russian).

Kozlovskii, V. V., Lomasov, V. N., and Vlasenko, L. S. (1988). *Radiat. Eff.* 105, 37–45.

Kozlovskii, V. V. and Abrosimova, V. N. (1991). Preprint. Inst. of Problems in Technology of Electronics and Ultrapure Materials, Inst. of Physics of Semiconductors, Siberian Division Acad. of Sciences of USSR, Chernogolovka.

Kozlovskii, V. V., Zakharenkov, L. F., and Shustrov, B. A. (1992). *Sov. Phys. Semicond.* 26, 1–11.

Kozlovskii, V. V. and Zakharenkov, L. F. (1994). *Petersburg J. Microelectr.* 1, 15–29 (in Russian).

Kozlovskii, V. V. and Zakharenkov, L. F. (1995). *Microelectr. J.* 26, 69–76.

Lang, D. V. (1977). In Proc. Int. Conf. on Radiation Effects in Semiconductors, Dubrovnik, 1976, pp. 70–94. Inst. of Physics, London.

Lark-Horowitz, K., Bleuler, E., Davis, R. E., and Tendam, D. L. (1948). *Phys. Rev.* 73, 1256–1264.

Lark-Horowitz, K. ([1950] 1975). In *Proc. Int. Conf. on Semi-conducting Materials*, Reading, England, 1950, H. K. Henisch, ed., pp. 47–51. Butterworths, London, and Academic Press, New York.

Maples, C., Goth, G. W., and Cerny, J. (1967). *Nuclear Data* A2, 3–248.

Meese, J. M., ed. (1979). *Neutron Transmutation Doping of Semiconductors. Proc. 2nd Int. Conf.*, Columbia, MO, 1978, pp. 214–240. Plenum Press, New York.

Mirianashvili, Sh. M., Nanobashvili, D. I., and Razmadze, Z. G. (1966). *Sov. Phys. Sol. St.* 7, 2877–2878.

Mirianashvili, Sh. M., and Nanobashvili, D. I. (1970). *Sov. Phys. Semicond.* 4, 1612–1613.

Mokhov, E. N., Didik, V. A., Gornushkina, E. D., and Kozlovskii, V. V. (1992). *Sov. Phys. Sol. St.* 34, 1043–1045.

Munzel, H. and Lange, J. (1969). In *Uses of Cyclotrons in Chemistry, Metallurgy and Biology. Proc. Conf.*, Oxford, 1969, Amphlett, C. B., ed., pp. 373–380. London.

Mushnikov, V. N., Khizhnyak, N. A., Shilyaev, B. A., et al. (1988). *Vopr. At. Nauki.* (Topics in Atomic Science and Technology.) Ser. Physics of Radiation Damage and Radiation Materials Science 5, p. 11 (in Russian).

Nemetz, O. F. and Gofman, Yu. V. (1975). In *Handbook on Nuclear Physics*, pp. 217–234. Naukova Dumka, Kiev (in Russian).

Prussin, S. and Cleland, J. W. (1978). *J. Electrochem. Soc.* 125, 350–352.

Smirnov, L. S. (1977). In *Physical Processes in Irradiated Semiconductors*, pp. 171–174. Nauka, Novosibirsk (in Russian).

Smirnov, L. S. (1980). In *Problems in Radiation Technology of Semiconductors*, pp. 204–211. Nauka, Novosibirsk (in Russian).

Smirnov, L. S. (1981). In *Doping of Semiconductors by the Nuclear Reaction Method*, pp. 17–31. Nauka, Novosibirsk (in Russian).

Trey, F. and Oberhauser, F. (1957). *Naturwissenschften* 44, 256–257.

Varnina, V. I., Groza, A. A., Kuts, V. I., et al. (1988). In *Proc. 6th All-Union Conf. on Physico-Chemical Basis of Doping of Semiconductor Materials*, pp. 168–169. Nauka, Moscow (in Russian).

Vavilov, V. S., Vodop'yanov, L. K., and Kurdiani, N. I. (1976). In *Radiation Physics of Nonmetallic Crystals*, pp. 206–212. Naukova Dumka, Kiev (in Russian).

Vavilov, V. S., Kekelidze, N. P., and Sirnov, L. S. (1988). In *Radiation Effect upon Semiconductors*, pp. 116–124. Nauka, Moscow (in Russian).

Vodop'yanov, L. K. and Kurdiani, N. I. (1966). *Sov. Phys. Sol. St.* 8, 55–56.

Zablotskii, V. V., Ivanov, N. A., Kosmach, V. F., et al. (1986). *Sov. Phys. Semicond.* 20, 397–398.

Zabrodin, B. V., Zakharenkov, L. F., Kozlovskii, V. V., and Shustrov, B. A. (1990). *At. Energ.* 68, 432–434.

Zakharenkov, L. F., Kozlovskii, V. V., and Shustrov, B. A. (1990). *Phys. St. Sol.* A117, 85–90.

Zatolokin, B. V., Konstantinov, I. O., and Krasnov, N. N. (1976). *Int. J. Appl. Radiat. Isot.* 27, 159–161.

CHAPTER 3

POLYMER DIFFUSANTS IN SEMICONDUCTOR TECHNOLOGY

E. G. GUK, A. V. KAMANIN, N. M. SHMIDT, V. B. SHUMAN, and T. A. YURRE

Impurity diffusion has become a basic method of obtaining silicon p-n junction device fabrication. At present there are two main diffusion methods that allow us to obtain a uniform and controlled surface concentration (N_S) of impurities with good reproducibility:

1. A diffusant is applied on the semiconductor substrate during the diffusion process.
2. A diffusant layer is formed on the semiconductor substrate before the diffusion.

The first approach requires expensive and sophisticated computer equipment. The equipment needed for the second approach is much simpler, since this approach does not require a uniform transfer of diffusant vapors to the substrate. One can obtain low surface concentrations in a one-stage diffusion process. Also this technique makes it possible to form a source of diffusion with

Semiconductor Technology: Processing and Novel Fabrication Techniques,
Edited by M. Levinshtein and M. Shur.
ISBN 0-471-12792-2

specified parameters directly on the semiconductor surface. The first spin-on diffusants were simply solutions of salts or acids. Then came published results by a number of authors who had tried to use film-forming compositions consisting of an organic polymer solution (e.g., cellulose) and solid oxides or some other chemical compounds containing dispersed doping elements (Lubashevskaya et al. 1968; Pat. USA 1969). However, since late 1960s a new method of a diffusion from silicone compositions has become widely accepted. This method allows one to form films that are sufficiently uniform and contain compounds of a doping element. Such films are formed as a result of preliminary chemical transformations of the silicone low-molecular reagents (frequently tetraethoxysylan) and chemical compounds of the doping element (Pat. USA 1963; Zee 1967). These ideas were further developed by others (Prikhid'ko et al. 1970; Pat. USA 1972), and their published results contain descriptions of compositions, including the catalytic agents of polycondensation that lead to solidification of the liquid film. Such improvements, however, have not eliminated the main drawback of compositions based on tetraethoxysylan—the poor stability of solutions due to the uncontrolled hydrolysis of tetraethoxysylan. The rate of hydrolytic polycondensation depends on such factors as the composition of the homogeneous medium and the absolute and relative content of components in the mixture. As a consequence, in order to obtain films of identical thickness, it is necessary to use solutions of the same age (Borisenko et al. 1970).

Diffusants based on tetraethoxysylan are still widely being used. The simplicity and relatively low cost of these processes make them competitive both with gas-phase diffusion and with ion implantation. Several authors (Ramamurthy 1987; Teh and Chuan 1989; Unger et al. 1990) have discussed the possibility of using boron and phosphorus containing spin-on dopants. However, the main drawbacks of these diffusion sources have not yet been eliminated: the above-mentioned instability, strong mechanical stress, causing the cracking of the diffusant film, and also the inevitable presence of SiO_2 in the composition of the film. The presence of SiO_2 complicates the impurity concentration profile. SiO_2 should also be removed even after the diffusion in an inert atmosphere.

There were continual attempts made to refine these materials in order to find a suitable substitute material for tetraethoxysylan. Yoldas (1980) has reported on the use of a clear polymerized solution derived from alkoxides of titanium and a p or n dopant.

It seems to us that the most radical method of eliminating the above-mentioned drawbacks would be to develop absolutely homogeneous polymer sources of diffusion that include the atoms of doping elements into their elementorganic components. This idea was put forward for the first time in our paper (El'zov et al. 1975). Thereafter some articles were published on the possibility of using polymers as a solid matrix for diffusant compositions. A number of papers have dealt with homogeneous solutions of film-forming silicone polymers and to an elementorganic compound of the elements of III and V groups (Beyer 1976, 1977). More recently there is described a method of diffusion from a film (Pat. USA 1993) that comprises an organic or inorganic linking compound and a doping element compound. Therefore it should be clear that the idea of using polymer compositions as a diffusion source has a long history. Generally, however, the polymer is suggested as a linking compound, providing a solid matrix in which the elementorganic compound or the doping element oxide can be distributed with a certain degree of uniformity and homogeneity.

We believe that chemical bonding of the doping element with the polymer can provide a more uniform distribution of the doping element in the polymer film. Such chemical bonding appears to be the result of a photostructuration process, or it may occur if an adequate elementorganic polymer is used. Then an exact dosage of the doping impurity can be achieved by choosing an elementorganic compound and varying its concentration in the composition.

It has been shown that practically any doping impurity can be introduced into the composition of a polymer diffusant and that several doping elements can be introduced simultaneously. Polymer diffusants make it possible for both deep and shallow diffusion layers of high quality to form on lapped and polished silicon surfaces, epitaxial films, and polysilicon. These diffusion sources are also very well suited for application in III-V compounds technology.

The diffusion sources retain stability for a period of six months to one year, depending on the composition. The diffusants are able to preserve the initial high lifetime of the minority carriers in silicon and also to produce an effective gettering of deep level impurities. By using the polymer diffusants, we have for the first time been able to accomplish both deep level impurity diffusion from a limited source and a simultaneous diffusion of deep and shallow impurities (Guk et al. 1986).

The high uniformity of diffusion layers has further made it possible to create diodes of large areas ($\geq 1\ cm^2$) with quasi-uniform avalanche breakdown at low current densities (Zubrilov and Shuman 1987; Zubrilov et al. 1989). The prospect of using these diffusants for the manufacture of solar cells has been demonstrated as well (Guk et al. 1995).

3.1. COMPOSITIONS OF POLYMER DIFFUSANTS

In the creation of polymer diffusants, azidecontent photoresists show the most promise as polymer diffusion bases (El'zov et al. 1975). Their main components are a photosensitive low-molecular arylazide and a film-forming polymer capable of structuring. Under the ultraviolet irradiation a photolysis of arylazide takes place with the formation of highly reactive half-products—nitrenes (stage 1)

$$\underset{\text{(Arylazide)}}{N_3{-}R{-}D{-}R{-}N_3} \longrightarrow \underset{\text{(Nitrene)}}{:N{-}R{-}D{-}R{-}N:}$$

Here D is a diffusion element, R is an organic radical, N is nitrogen.

These compounds cause transverse crosslinking of polymer chains which results in the formation of a photorelief with predetermined parameters (stage 2)

~ P−D− ··· −P− ··· −D−P ~
(Polymer chain)
+ N−R−D−R−D ⟶
~ P−D− ··· −P−P− ··· −D−P ~
+ N−R−D−R−N ⟶
~ P−D− ··· −P−P− ··· −D−P ~

~ P−D— ··· —P−P— ··· —D−P ~
N−R−D−R−N
~ P−D— ··· —P−P— ··· —D−P ~
N−R−D−R−N
~ P−D— ··· —P−P— ··· —D−P ~

Here P is part of a polymer chain.

The use of elementorganic compounds as photosensitive or polymer components makes it possible to form polymer layers that incorporate the atoms of the doping element, directly in the composition of the hard cured network structure. The doping atoms are chemically bound and uniformly distributed in the polymer layers. Using such polymer layers as sources of diffusants allows one to control the content of the doping impurities. It is easy to obtain a low concentration of the diffusion impurities by introducing the doping element into a low-molecular photostructuring component, whose content in the composition is relatively small (5–10%).

Highly doped diffusion layers are obtained using compositions containing a doping element either in the polymer or in the both components. This way it is possible to control the surface concentration N_s by diluting the composition either with a solvent or with an organic polymer that does not contain any dopants.

Specific requirements to every component are determined by the requirement that organic polymer diffusants form films that are 0.1–1 μm thick. The polymer base of the diffusant must be able to form an uniformally strong and solid film with good adhesion to the semiconductor substrate surface. The autocohesion of the film-forming polymer must be high in order for the film thickness to vary. In the formation process of the polymer diffusant film, the system of solvents is especially important. It determines both the interior stresses of the thin films and the surface tension value. The low values of surface tension provide a small deviation of the predetermined film thickness. The necessary condition, for creating compositions, is an excellent mutual compatibility and solubility of the low-molecular and polymer components. An important factor is also the stability of the components and the absence of undesirable impurities. All these requirements must be taken into consideration when forming a wide set of polymer diffusants.

3.2. POLYMER DIFFUSANT LAYER FORMATION

To obtain the polymer diffusant layers of good quality, all the basic conditions for film-forming components, elaborated in the process of using ordinary photoresists, have to be considered as well: an optimal viscosity of solutions, solvent volatility, film-forming properties, and good adhesion of polymers. Hence the rules for using polymer compositions coincide to a great extent with those for ordinary photoresists discussed by Moreau (1988).

First, the solution of components must not be near saturation; otherwise, immediately after the wafer has been covered with drops of solution, the deposition may occur, and the deposited particles will affect the film uniformity just like foreign contaminations.

Second, layers of polymer diffusant on the wafers may be affected by thermodestruction in the oxidizing atmosphere. During thermodestruction the organic compounds decompose, and the doping element oxides form a uniform film on the semiconductor wafer surface.

Third, to obtain a uniform oxide film, it is necessary to choose the optimal time and temperature of the process.

The choice of the optimal thermodestruction temperature is conducted on the basis of a thorough study of the process. As an example of such an investigation for a sufficiently thermostable polymer with a high content of boron, consider the polycarborane esterified adipinic acid

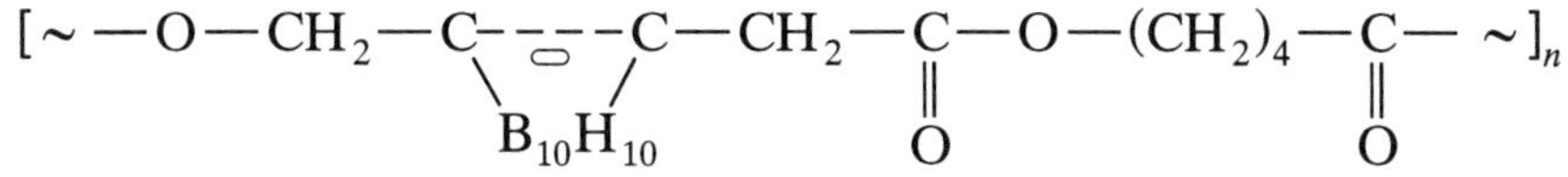

$$[\sim -O-CH_2-\underset{\diagdown\ \ \ \ \diagup \atop B_{10}H_{10}}{C\overset{o}{-\!-\!-}C}-CH_2-\underset{\| \atop O}{C}-O-(CH_2)_4-\underset{\| \atop O}{C}- \sim]_n$$

Due to a greater thermostability of carboranes compared to organic polymers as well as to photosensitive azides, the study of thermodestruction was conducted on the polycarborane without other composition components.

The derivatogram, made for the product under investigation (Fig. 3.1) enables one to observe the temperature-dependent polymer transformation and to estimate the lowest temperature at which the decomposition occurs. The estimation of the completeness of thermodestruction is made by comparing the IR spectra of the polymer film before and after annealing (Fig. 3.2). In the spectrum of the original film (see Fig. 3.2, curve 1) absorption bands characteristic of this polymer are observed: The 2962 and 2872 cm^{-1} bands correspond to the asymmetric and symmetric stretching vibrations of the CH_3-group, the 2926 and 2853 cm^{-1} bands correspond to the asymmetric and symmetric stretching

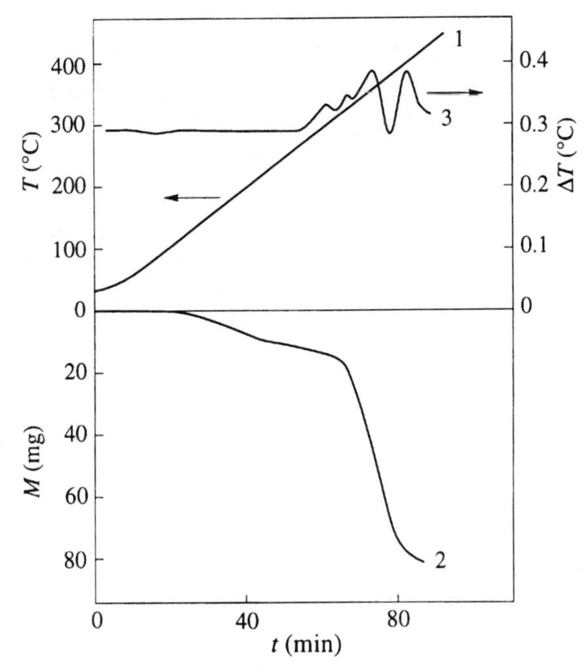

Figure 3.1. Derivatogram for polycarborane esterified adipinic acid: (1) Curve of heating; (2) thermogravimetric curve; (3) curve of differential thermal analysis.

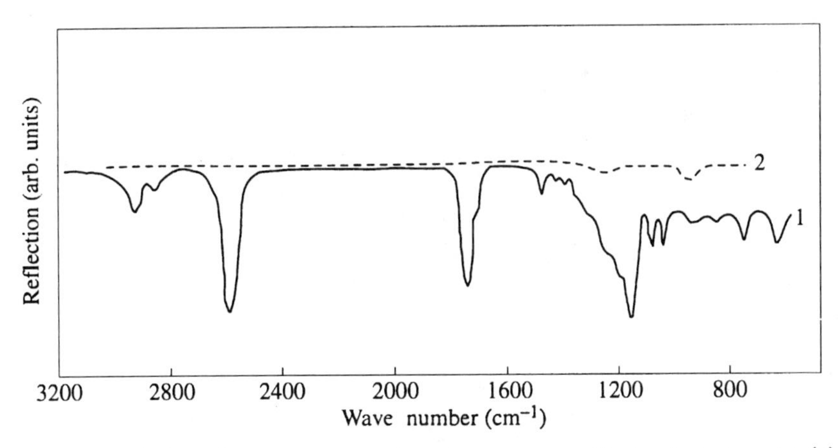

Figure 3.2. Infrared spectra of polycarborane esterified adipinic acid (1) and of the product obtained after its thermal destruction lasted for an hour at 450°C (2).

vibrations of CH_2-group, the 1460 and 1370–1380 cm^{-1} bands correspond to deformation vibrations of CH_2-group, and the 720–750 cm^{-1} bands correspond to the rock vibration of $(CH_2)_4$-group (Bellamy 1954). The intensive 1745 cm^{-1} band corresponds to the carbonyle stretching vibrations, and the wide, very intensive band with the maximum at 1150 cm^{-1} corresponds to the vibration of the C—O—C bond of the polyester group (Gordon and Ford 1972). In addition, in the original film, one observes a 2595 cm^{-1} band that is caused by the bond B—H contained in the carborane nucleus (Bellamy 1968). Under thermal treatment at 250°C for one hour, the spectrum did not change. After the thermodestruction at 450°C, which also lasted one hour, only a wide weak absorption band of B—O bond is seen near 1300 cm^{-1} (see Fig. 3.2, curve 2) (Welther and Warn 1962).

Thermodestruction at 450°C brings on decomposition of the polymer which oxidizes to B_2O_3 and remains on the semiconductor substrate surface, serving as a source of boron during diffusion. Thermodestruction thus raises the surface concentration and typically improves the impurity concentration's uniformity when the fusion temperature or at least the boiling point of diffusant oxides is higher than the destruction temperature.

After thermodestruction diffusion is carried on within the temperature range of 850°–1300°C as long as required.

3.3. III-V IMPURITIES IN SILICON

3.3.1. Diffusion of Boron, Phosphorus, Antimony, and Arsenic

Boron and phosphorus are the diffusion impurities widely used in silicon technology. The diffusion of boron and phosphorus from the polymer films into silicon in the air has been studied since 1977 (Guk et al. 1977).

To estimate the value of the surface concentration, one must know the impurity concentration profile in the diffusion layer. The amount of the doping element in the composition of the polymer diffusant applied on the surface of silicon varies within several orders of magnitude: from $\sim 10^{14}$ cm^{-2} to $\sim 10^{18}$ cm^{-2}. The experimentally found dependencies of the surface conductivity σ_s

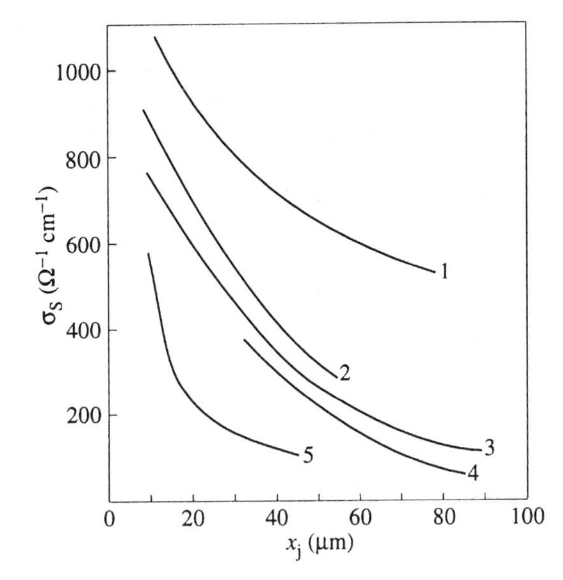

Figure 3.3. Dependence of surface conductivity (σ_s) on the depth of the p-n junction (x_j) with the diffusion of different boron concentrations (1–3% B, 3–10% B, 4—20% B) and the diffusion of different phosphorus concentrations (2–10% P, 5–20% P) in air at a 1250°C diffusion temperature.

on the depth of p-n junction x_j are indicated in Figure 3.3. The obtained results do not exactly fit either the erfc-function or the Gaussian function. Since diffusion is taking place in the oxidizing ambient, there is an impurity redistribution between silicon and SiO_2 layers. The impurity vaporization from the surface is also important. The departure from the Gauss's law is, however, quite small for boron and phosphorus, and in practice one can use the standard curves for N_s evaluation (Irvin 1962). The region of very high concentrations is the only exception. For the maximum concentration of the boron-containing polymer, the value of σ_s, calculated from the measured surface resistance and x_j value is larger than 1500 $\Omega^{-1} \cdot \mathrm{cm}^{-1}$, which corresponds to the surface concentration of more than 10^{21} cm^{-3}, a concentration that is higher than the solid solubility of boron in silicon. Similar results have been obtained for phosphorus. They can be explained by a characteristic departure of the actual diffusion profile in the region of maximal concentrations from both erfc-function and Gaussian function (Burger and Donovan 1967). However, the investigation of diffusion of antimony and arsenic from the polymer source confirms the proximity of the impurity distribution profile to the Gaussian function.

The surface concentration of B, P, Sb, and As for a particular diffusion regime depends linearly on the film thickness of the diffusant (Fig. 3.4). If both the polymer composition viscosity and the film thickness of the diffusant (d) are kept constant, the dependence of the surface concentration on the doping impurity content in the polymer diffusant (M) is a linear function of M in a wide interval of N_s (from 10^{17} to 10^{20} cm^{-3}); see Figure 3.5. Figures 3.4 and 3.5 show that for the given diffusion regimes of B, P, Sb, and As the value of N_s is determined solely by the amount of the impurity spin-on in the composition of the polymer diffusant per unit of the wafer surface. At the same time, for the tetraetoxysylan based diffusants, the dependence of the surface impurity concentration is nonlinear (Borisenko et al. 1972), which causes additional complications.

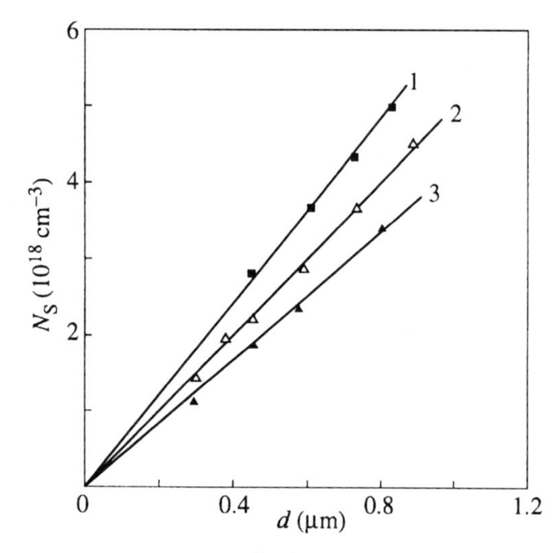

Figure 3.4. Surface concentration (N_s) dependence of boron (1), phosphorus (2), antimony, and arsenic (3) on the film thickness (d) of a polymer diffusant.

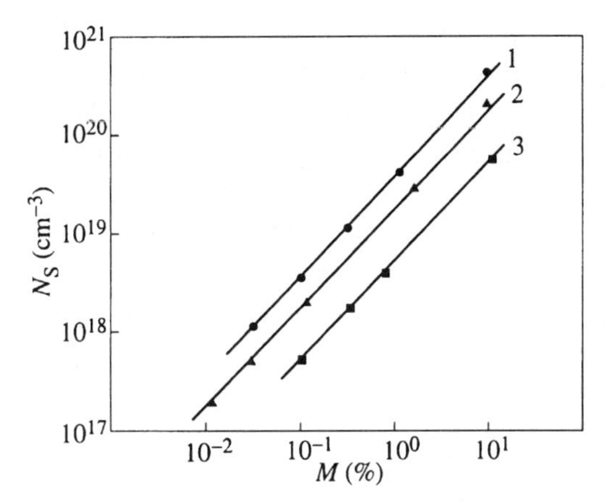

Figure 3.5. Dependence of the surface concentration (N_s) of boron (1), phosphorus (2), antimony, and arsenic (3) on the content (M) of the doping element in the composition, the film thickness of the diffusant being $d = 0.5$ μm.

For boron and phosphorus diffusion from the polymer sources in the inert atmosphere (for the same diffusion regimes), the value of N_s is slightly higher due to the absence of the impurity redistribution between the silicon and the oxide layer. However, all important features of the process described above remain basically the same.

The deviation of the values of N_s on the wafers with a diameter of 60 mm varies between $\pm 5\%$ for low and $\pm 3\%$ for high values of the surface impurity concentration in the diffusion layer for any diffusion regime. Thus we succeeded in achieving a uniform and reproducible introduction of the group III and V impurities into the silicon. The obtained results are comparable to those obtained by other widely used methods (e.g., diffusion from a gaseous phase, diffusion from spin-on sources on the tetraethoxysylan base, and ion implantation). However, this method is simpler, more convenient, and cheaper.

3.3.2. Diffusion of Aluminium

The diffusion of aluminium differs in a number of ways from B and P diffusion. For example, unlike boron which can be reduced from

B_2O_3, Al cannot be reduced by silicon from a higher oxide. For diffusion from a spin-on source in the oxidizing atmosphere, the distribution behavior of Al in a diffusion layer is always in accordance with Gaussian law in that the diffusion occurs from a limited source (Grekhov et al. 1966). This behavior is characteristic of aluminium as a chemical element irrespective of the nature of the diffusion source that is being applied. We studied the diffusion of Al in the air into the lapped n-type silicon wafers ($\rho_0 = 100$ $\Omega \cdot$ cm) from a polymer diffusant. Figure 3.6 shows the dependence of σ_s on the time of Al diffusion at 1250°C. The dependence of the surface concentration of Al on its content in the diffusant is shown in Figure 3.7. It can be seen that when the atomic fraction of Al exceeds 10^{-2}%, its surface concentration reaches 10^{17} cm^{-3} (which is its maximum value with the diffusion taking place in the oxidizing atmosphere).

For the aluminium diffusion into a polished surface under the same conditions, the concentration of Al becomes lower by more than an order as in case of using other spin-on sources of that element.

Conducting the diffusion of Al in an inert gas atmosphere greatly raises N_s because of the absence of SiO_2 on the silicon surface and the higher vapor pressure of Al_2O in the non-oxidizing atmosphere (the reaction $Al_2O_3 \Leftrightarrow Al_2O + O_2$ in the inert atmosphere is shifted to the right). However, the value of N_s even in an inert

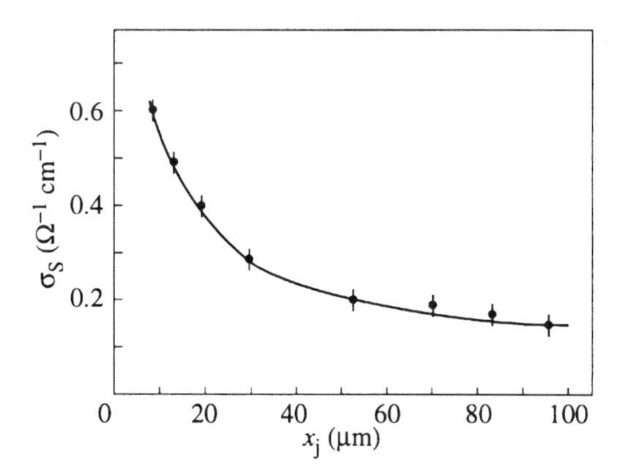

Figure 3.6. Dependence of the surface conductivity (σ_s) on the depth of the p-n junction (x_j).

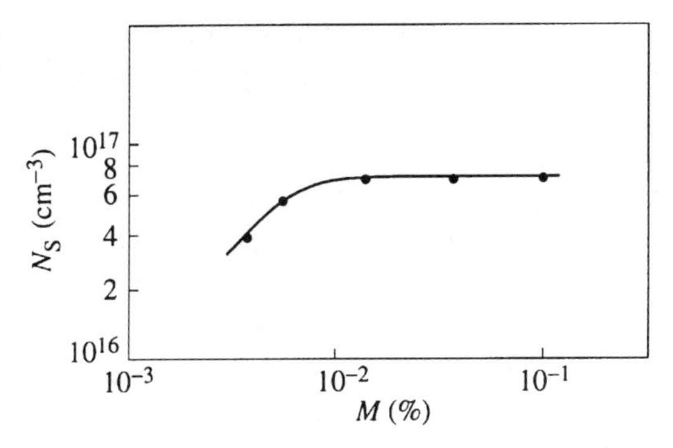

Figure 3.7. Dependence of the surface concentration (N_s) of aluminium on its content (M) in the composition.

atmosphere remains lower (by approximately an order of magnitude) than the solid solubility of Al in silicon which at the temperature of diffusion is about $2 \cdot 10^{19}$ cm^{-3}. As was mentioned above, in contrast, for boron and phosphorus diffusion from polymer sources, the solid solubility is achieved quite easily. However, just like for the boron and phosphorus diffusion, it is easy to control the concentration of aluminium. A high distribution uniformity of the Al concentration on the wafer surface has been achieved ($\pm 3\%$) as well as a good reproducibility of these results from wafer to wafer and from batch to batch. These results, achieved by a simple and efficient method, enable us to use the Al diffusion for device fabrication.

3.4. DIFFUSION OF DEEP-LEVEL IMPURITIES IN SILICON

We have given examples of polymer diffusants containing elements of groups III and V. In a similar way polymer compositions containing gold, platinum, and sulphurorganic compounds or complexes of these elements with organic ligands have been created.

3.4.1. Diffusion of Gold

Introducing gold into silicon has been used for a long time to obtain a high-resistance compensated material and to decrease the minority carrier lifetime. It is also possible to control the minority carrier lifetime by introducing radiation damage into silicon. But the radiation-introduced defects are annealed very soon at temperatures above 350°C.

The common method of doping silicon with gold is a diffusion from an unlimited source (from either sputtered or chemically deposited layer of Au or from its salt); (see Sprokel and Fairfield 1965; Martin et al. 1966). As this takes place, the concentration of gold in the volume of silicon wafer N_{Au} can be regulated just by choosing a suitable regime of diffusion. It is necessary to maintain the temperature of diffusion (from 850°C and higher) within 1°C and the time of diffusion with an accuracy of several minutes. But even these tolerances do not guarantee reproducible diffusion, since the concentration of the diffused Au depends on the concentration of shallow impurities. When the temperature of the diffusion is below 1000°C, the concentration of Au depends also on the dislocation density in silicon and on the preceding thermal treatment of the wafers (Badalov and Shuman 1969). Besides, with a diffusion from an unlimited source, the gold concentration profile has a characteristic U-shape. It means that even with a one-lateral application of the diffusant, the gold concentration has a maximal value in a subsurface layer, dropping smoothly with the increase of distance from the surface, coming to the plateau in several dozens of microns, and then very gradually rising again, reaching its maximum in the opposite subsurface layer. At the plateau the value of N_{Au} is proportional to $t^{1/2}$, where t is the diffusion time (Sprokel and Fairfield 1965; Martin et al. 1966). The deviation of the gold concentration across the wafer reaches 100%. Such a behavior of gold in silicon was first explained by the dissociative mechanism of diffusion (Wilcox and La Chapelle 1964). The results were interpreted in the following way: Apart from the diffusion of gold via interstitials and its transition from the interstitials into the vacancies, there is a diffusion flow of vacancies both from the surface and from structure defects. Thus constant inflow of vacancies determines the rate of gold transition from the interstitials into vacancies until the solid solubility is reached. If the diffusion takes place at the temperature below 1000°C, it may be affected by vacancies and structure defects present in the original material.

The application of a polymer, containing an elementorganic compound of gold, makes it possible to apply the controlled amounts of gold uniformly on the silicon surface and to carry out the diffusion from a *limited source*, namely by using the same technique as when introducing the impurities of groups III and V.

Guk et al. (1982) have studied the diffusion of gold from the limited source into the n-Si wafers 0.35-mm thick with the initial resistivity (ρ_0) equal to 5.5–100 $\Omega \cdot$ cm. To estimate the effect of the original silicon dislocation density on the results of the diffusion, they used wafers with the dislocation densities of 10 and 10^4 cm^{-2}. The wafers were first etched in KOH. The diffusion was carried out in the air within the temperature range of 1000° to 1200°C. The diffusion time varied from 15 minutes to 20 hours. In order to study the effect of the cooling rate, two different regimes of cooling were used: the air quenching and the cooling to 800°C at the rate of 1°C per minute.

After the diffusion the value N_{Au} was determined from the degree-of-compensation ratio provided that $0.3N < N_{Au} < 1.5N$, where N is the shallow donor concentration. A high uniformity of the gold concentration in the volume of the wafer was achieved: the N_{Au} profile did not have a U-shape. The deviation of the gold concentration, measured using lapping, did not exceed 5% across the thickness of the wafer and across its area. This method allows one to obtain reproducible results irrespectively of the sample history: neither the dislocation density nor the shallow impurity concentration affects the results. The value of N_{Au} is not affected by the rate of sample cooling after the diffusion, and when the diffusion temperature is above 1000°C, the gold concentration does not depend on the time of diffusion either (when it changes from 15 minutes to 10 hours and more). The value of N_{Au} can be controlled to a sufficient degree of precision by the quantity of the gold compound applied to the surface of wafers before the diffusion.

Despite the seeming contradiction with the dissociative mechanism, these results can be explained within the framework of this theory. When the diffusion is carried out from a limited source, there is no constant inflow of Au. The source will get exhausted very soon, since not more than 10^{-2} atom layers of gold had been applied on the silicon surface. Since the diffusion temperature selected is high, the formation of gold complexes practically does not affect diffusion (Badalov and Shuman 1969; Badalov and Shuman 1970), and the effective gold diffusivity D remains high even in the dislocation-free silicon. Thanks to a high interstitial diffusivity (10^{-5}–10^{-6} $cm^2 \cdot s^{-1}$), the gold atoms fill the volume of the wafer during time $t \sim W^2/8D \simeq 2$ to 3 minutes, where W is the thickness of the wafer. The vacancy concentration in the original silicon can be of the same order as the concentration of diffused gold or larger. Consequently the rate of the transition of Au into lattice sites is determined not by the inflow of vacancies but by the diffusion of Au via interstitials. Therefore the process of the transition of Au into vacancies soon comes to an end: In diffusion from a limited source for diffusion time $t > 15$ minutes, $N_{Au} \simeq$ const. is observed; near the surface, where the vacancy concentration is higher, there is practically no layer with a higher concentration of Au. The fact that N_{Au} is constant even if diffusion lasts long is a proof that Au is not released in appreciable quantities on the $Si-SiO_2$ interface or on any other structure defects.

We thus managed to obtain N_{Au} in the range of 10^{13} to 10^{15} cm^{-3} with a good reproducibility (Guk et al. 1982). The deviation of the value of N_{Au} on the surface of the wafer (25–30 mm in diameter) was not more than 5%. The rate of cooling did not affect the results; at the final temperature of slow cooling, 800–850°C, the solution of gold in silicon in the above-mentioned range of concentrations did not become supersaturated (the solid solubility of gold in silicon at 850°C being 10^{15} cm^{-3}); (see Wilcox and La Chapelle 1964).

In the case of a polymer diffusant with a high content of aurumorganic compounds, it is possible to introduce into silicon higher concentrations of gold until it reaches its solid solubility in silicon at the diffusion temperature. In this case, however, the samples must be quenched after the diffusion in order to avoid the supersaturation of the gold solution in silicon, which would result in forming a solid phase and in a decrease of the gold concentration.

Thus the introduction of Au into silicon by diffusion from a limited source has that advantage that N_{Au} (at $W =$ const.) depends on only one parameter—the quantity of the applied compound of Au. In contrast, in the standard case of the diffusion from an unlimited source, N_{Au} depends on many parameters.

It should be noted that polymer diffusants allow one to obtain a maximum concentration of Au, which is a limit of its solubility in silicon. In this case the polymer composition contains a large amount of the aurumorganic compound and may serve as an unlimited source for the diffusion.

3.4.2. Diffusion of Platinum

The control of the minority carrier lifetime using gold is inevitably accompanied by compensation, since gold creates a deep level in the middle of the silicon band gap (Bullis 1966). Often such a compensation cannot be tolerated. Moreover the deep level's location in the middle of the band gap adversely affects the characteristics of rectifier structures, raising the generation current and also reverse leakage. The use of platinum, which creates deep levels far enough from the middle of the band gap $E_c - 0.24$ eV, $E_v + 0.34$ eV, (Conty and Panchiery 1971), may help avoid this problem. Having an effective recombination center, platinum provides a better characteristics of silicon p-n junctions. A solution of $HPtCl_4$ was used in introducing platinum, and the diffusion, just as in the case of gold, was performed from an unlimited source (Carhano 1970). All the drawbacks typical of this method and described above (see Section 3.4.1) were also observed for Pt diffusion. However, the diffusion of Pt revealed a number of additional peculiarities. We should note that different authors observed the dispersion by more than an order in the minority carrier lifetime, when Pt diffusion was performed in the same regime (Lisiak and Milns 1975; Mielke 1975).

We carried out the diffusion of platinum from a limited source, using a polymer diffusant containing platinum. In order to evaluate the uniformity of the Pt distribution, we used the p^+n epitaxial films and fabricated the p^+n diode structures using boron diffusion into n—Si with resistivity of 2 $\Omega \cdot$ cm. The initial hole lifetime exceeded 5 μs. After spinning a polymer and applying the thermodestruction at 450°C for 30 minutes, diffusion was achieved in air at temperature of 900°–1200°C. After the diffusion of Pt and the ohmic contact fabrication, the wafers were divided into 2×2 mm structures for the hole lifetime (τ_p) and the reverse recovery time

(τ_o) measurements. The adjustment of Pt concentration made it possible to obtain τ_p from 2 μs to 5 ns (in diodes this time was 1.5 times smaller than in epitaxial structures). With the maximum content of platinum in the composition, the concentration close to its solid solubility in silicon was achieved. For the maximum concentration of platinum at the diffusion temperatures of 1100°–1200°C, the reverse recovery time did not depend on the time of diffusion, ranging from 40 minutes to up to 20 hours. Cooling the samples after the diffusion at 1°C per minute to 800°C reduced the concentration of platinum compared to the regime of an abrupt cooling (quenching). Lowering the diffusion temperature to 900°C led to the rise of τ_o because of the decreased solubility of platinum in silicon.

After the introduction of platinum, a diffusion of phosphorus for forming the p^+nn^+ structures was performed. The diffusion of phosphorus did not lead to the gettering of platinum if it was accompanied by quenching. This result confirms that Pt has a much weaker interaction with the n^+ layer than Au (Mielke 1975).

The uniformity of the distribution of the hole lifetime (and accordingly of the reverse recovery time) for all of the samples depended very much on the quality of silicon and epitaxy films, namely on the presence of structural defects and additional impurities. But the deviation of τ_o in the best samples did not exceed 10%. Nevertheless, the deviation of the values in the worst samples and from sample to sample could reach 50% for the same diffusion regime. This shows that the diffusion of platinum from the polymer composition allows one to eliminate the drawbacks of the traditional diffusion. However, other complications mentioned above indicate the peculiarities of the interaction of silicon and platinum. One can assume that a certain fraction of the Pt atoms can be electrically inactive. On the other hand, both the Pt ion and a complex of Pt can serve as a center of recombination. The formation and the behavior of such a center depends on the presence of structural defects in a silicon wafer.

3.4.3. Diffusion of Sulphur

With the exception of oxygen, the behavior of nonmetallic impurities in silicon has not been examined in detail. In particular, the influence of such an impurity as sulphur on the characteristics of silicon devices has been hardly examined at all, even though sulphur creates in silicon donor levels at $E_c - 0.18$ eV and $E_c - 0.37$ eV and has diffusivity that is three orders higher than aluminium (Carlson et al. 1959). An investigation of sulphur's effect on silicon seems to be of interest for two reasons: (1) for possible controlled use of sulphur during device fabrication (e.g., for long-wave photodetectors) and (2) for better knowledge of inadvertent sulphur contamination on device parameters. Sulphur is one of those unintentional impurities introduced both from positive photoresists which contain sulphur and from atmospheric dust which harbors a large amount of sulfide compounds. Many technologists believe that due to its high volatility, sulphur cannot get into diffusion layers and consequently does not affect device performance. We have shown (Guk et al. 1985) that, in fact, even a negligible amount of sulphur (just a few atomic layers) on silicon before the thermal treatment can cause a drop of the breakdown voltage in silicon devices and lead to a thermal instability of silicon, such as has been formerly explained solely by the complicated behavior of oxygen in silicon (Kaiser and Keck 1957).

A polymer film containing sulphazide applied to a diode structures or deposited on the n-type silicon wafers ($\rho_0 = 20$–$200\ \Omega \cdot$cm) and p-type wafers ($\rho_0 = 10$–$700\ \Omega \cdot$cm) can serve as a source of sulphur. The use of a polymer source enables one to regulate the quantity of sulphur per unit of wafer area over a wide range, and that eliminates the erosion of the silicon surface, which is inevitable when a sulphur diffusion takes place in a sealed ampoule (Carlson et al. 1959). A high-quality surface is retained even at a maximal quantity of the diffusant $\sim 10^{17}$ cm^{-2}, namely up to $\sim 10^2$ atomic layers of sulphur, while the concentration of sulphur in the formed diffusion layer is close to the solid solubility. Due to an extremely high volatility of SO_2, no preliminary thermodestruction of the polymer film is performed as the sulphur diffusion is taking place. The diffusion process occurs in air at temperatures of 1000°–1200°C, with the process time varying from 15 minutes to several hours. Two regimes of cooling are used, the rate being either $\geq$ 100°C per minute or 1°C per minute.

After the surface layer has been removed, the type of conductivity and the value of resistivity on the wafers allow one to determine the concentration of the electrically active sulphur. As a result of the sulphur diffusion, the p-type silicon is transformed into an n-type silicon, and in the n-type Si the value ρ is reduced. The variation of ρ across the wafer is fairly small ($< 5\%$). The minimum value of ρ, obtained as a result of diffusion at 1230°C is 2–2.5 $\Omega \cdot$ cm. This reveals the presence of $\sim 2 \cdot 10^{15}$ cm^{-3} of sulphur, which is close to the value of its solid solubility in Si (10^{16} cm^{-3}) at this temperature (Carlson 1959). The use of a less concentrated source can lower the concentration of S by an order of magnitude. With the increase of the diffusion time from 15 minutes to 1 hour, the value of ρ decreases to a certain extent, but after that it practically does not change at all. This agrees with the estimation of time necessary to uniformly saturate a thin wafer with an impurity in a given temperature range. The concentration of sulphur practically does not depend on the rate of cooling or on the concentration of dislocations (in the range $0\text{–}2 \cdot 10^4$ cm^{-3}), nor on the type of conductivity or on the resistivity of the original silicon. All the other conditions being the same, the sulphur concentration depends on the surface treatment of the original wafers: In a lapped silicon (i.e., in silicon with a deep damaged layer) it is about an order higher than in etched silicon. This proves that the silicon surface traps sulphur only during the initial stage of thermal treatment, just as in the case of Al diffusion in an oxidizing atmosphere. Thus we can state that sulphur diffusion from a limited source can take place.

Sulphur penetrates through a p^+-layer rather easily. At the base of p^+np^+ structures the concentration of sulphur appears to be only several times lower than in the standard wafers without the p^+-layer. At the bases of n^+pn^+ and n^+nn^+ structures, the sulphur concentration is two orders lower than in the standard wafers. This shows that an n^+-layer retards sulphur. The gettering of sulphur, introduced onto silicon by a subsequent diffusion of phosphorus with a surface concentration $\geq 10^{20}$ cm^{-3}, has a low efficiency. (After a four-hour diffusion of phosphorus at 1230°C, the concentration of S in a 0.35-mm-thick wafer was reduced only thrice.) The long time annealing of wafers with sulphur at 800°C lowers the sulphur concentration very little. In the latter case the gettering proceeds very slowly at the expense of the decay of the solid solution, since the diffusivity of sulphur at 800°C is quite small

($< 10^{-10}$ $cm^2 \cdot s^{-1}$). The sulphur diffusion reduces breakdown voltages (U_{br}) of diode structures. After diffusion these voltages correspond to the value of the resistivity of the base. For example, if before the diffusion of sulphur the breakdown voltage was $U_{br} = 10^3$ V, after the diffusion of sulphur (sulphur concentration $\sim 1.5 \cdot 10^{15}$ cm^{-3}), the breakdown voltage is $U_{br} = 250$ V.

The low efficiency of sulphur gettering could be used to an advantage in diffusion of this element into silicon for device fabrication. However, it is important that any trace of sulphur be avoided in high-voltage devices.

3.5. JOINT DIFFUSION OF IMPURITIES IN SILICON

The chemical compatibility of different elementorganic compounds makes it possible to create polymer compositions containing simultaneously several doping elements. Homogeneous films of these compounds allow one to carry out a joint diffusion of several impurities simultaneously. The use of such mixed polymer diffusants allows us to predetermine and regulate the contents of separate elements and their relationships in film. Using the mixed sources of diffusion, we have succeeded in realizing a simultaneous diffusion of shallow impurities and a joint diffusion of shallow and deep impurities. The latter was possible only in using the diffusion of deep impurities from a limited source (Guk et al. 1982). Below we will give a number of examples of successful joint diffusion.

3.5.1. Simultaneous Diffusion of Phosphorus and Antimony

A high phosphorus concentration in a diffusion layer leads to dislocations in silicon. This phenomenon is caused by the difference in the atomic radii of the doping element and the atoms of the lattice. The introduction of such atoms into the silicon crystal creates a mechanical stress, resulting in plastic deformation and subsequent dislocations (Kurnosov and Yudin 1974). The presence of dislocations in a diffusion layer leads in turn to an abnormal impurity distribution, which tells quite negatively on the operation of the devices. It is known, however (Jagi et al. 1970; Fujimoto and

Komaki 1972), that a simultaneous doping of silicon with two impurities, one of which (phosphorus) has a smaller atomic radius (1.10 Å) than that for silicon (1.17 Å), and the other one (arsenic, antimony) a larger one (1.36 Å), leads to an abrupt reduction of the dislocation density in the diffusion layer. Then there is a mutual compensation of stresses caused by the impurities. It has been shown (Koledov and Popova 1973) that the optimal relation of phosphorus and antimony concentrations in the diffusion layer is Sb : P = 1 : 3.3. To obtain a dislocation-free layer it is sufficient to have the ratio as low as 1 : 15: that is, it is enough just to partially compensate the stresses that appear when silicon is doped by phosphorus.

By using elementorganic compounds, it is possible to obtain mixed compositions containing phosphorus and antimony (arsenic) in a desired proportion. Phosphorus can be contained in heterochain polymer, and arsenic or antimony can be contained in photostructuring components. The presence of one in polymer and the other in arylazide, both being completely compatible with each other, results in their uniform distribution in a polymer film. That has enabled us to obtain diffusion layers whose dislocation density is by five orders lower than that in the layers produced by phosphorus alone (10^3 and 10^8 cm^{-2}, respectively; Guk et al. 1984).

3.5.2. Joint Diffusion of Boron and Aluminium

Joint diffusion of boron and aluminium is widely used for diffusion layers in multilayer structures, in particular, in obtaining the p-base of the thyristors. Boron provides the required surface concentration ($\sim 10^{18}$ cm^{-3}), and aluminium, which has larger diffusivity, results in a small gradient of impurity in the p-n junction. Note that the surface concentration 10^{18} cm^{-3} is formed by only about two atomic layers of boron that have diffused into the bulk. Therefore obtaining a low concentration of boron is a challenge. With all the other conditions being alike, the reproducibility of the results depends on seemingly secondary details of the process: during diffusion whether the wafers are piled or are placed separately one by one in the slot of a quartz holder, whether there is large gas flow, whether vapor is present, and so on. These are some of the factors that affect the vaporization of the impurity and its transformation into the oxide, therefore affecting the surface concentration value.

Polycarborane—one of the polymers containing boron—has been used in such compositions. Complexes containing aluminium and boron, which have been used as photostructuring agents, are compatible with this polymer. The optimal content of the components in the composition was determined by the compatibility of the azide and the polymer. The homogeneity of the system and also the maximum solubility of the azide in the polymer and in the solvent were taken into account. The application of those mixed sources of diffusion enabled a high reproducibility of diffusion layers with the predetermined value of the surface boron concentration (its deviation did not exceed 3% over the area of the wafer). At the same time the desired depth of the p-n junction was determined by aluminium, whose diffusion coefficient is higher by an order of magnitude. A high breakdown voltage was achieved because of a small concentration gradient in the p-n junction. The value of the breakdown voltage corresponded to the theoretical value of the material with the given concentration of a shallow impurity calculated by Van Osterstraeten and De Man (1970).

3.5.3. Join Diffusion of Boron and Platinum

In the traditional approach to the formation of high frequency rectifiers, first a silicon structure is formed with the p-n junctions, and then gold or platinum is introduced at a lower temperature (800°–900°C) in order to create the necessary concentration of the recombination centers in the base. The diffusion of deep impurities from a limited source should allow one to carry out the joint diffusion of deep and shallow impurities at high temperatures (900°–1200°C). Making use of a good compatibility of the solutions of polymers containing Pt and B, we studied the possibility of a simultaneous diffusion of boron and platinum (Guk et al. 1987).

We used polished and lapped n–Si wafers with $\rho_0 = 0.5\ \Omega \cdot \text{cm}$. After forming a film of the polymer diffusant, comprising elementorganic compounds of boron and platinum, its destruction was performed for 30 minutes at 450°C. The subsequent diffusion took place in the air. After a one-side removal of the diffusion layer either a diffusion of phosphorus at 1140°C can be performed (from the adequate polymer film containing phosphorus) or nickel contacts applied.

The concentration and uniformity of the platinum distribution was estimated from the value and homogeneity of the hole lifetime (τ_p) in the diode base (in the standard samples, manufactured without Pt, $\tau_p > 1$ μs).

In addition to Pt concentration, several factors affect the value τ_p. For example, the method of the surface treatment of the wafer affects the value τ_p: All the other conditions being the same (e.g., when the diffusion lasts one hour at 1200°C), the value τ_p in the wafers with lapped surface is three to four times lower than in the wafers with a polished surface. Due to instability of a solid solution of Pt–Si (see Section 3.4.2), the rate of cooling after the diffusion also affects the results. For example, cooling at 800°C at the rate of 1°C per minute raises τ_p in several times. Therefore, as a rule, quenching was used, that is, cooling the wafers to room temperature in a few seconds. With the increase of diffusion time, the value τ_p decreased (Fig. 3.8). The concentration growth of platinum with the increase of the diffusion time may indicate that the diffusion is being limited by vacancy concentration. The vacancy concentration was defined both by the generation of the vacancies in the bulk of the material and by the inflow of vacancies from the surface. This

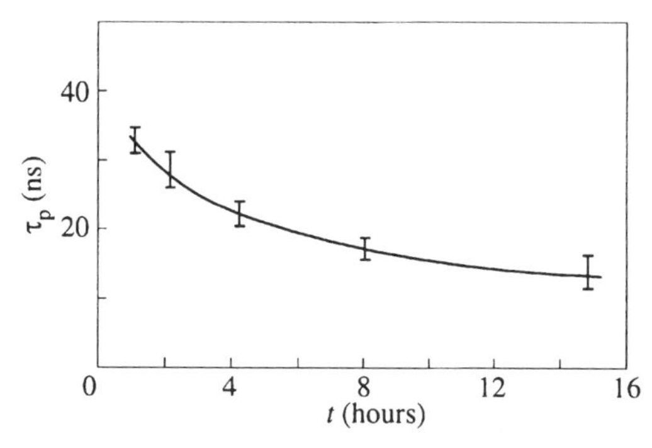

Figure 3.8. Dependence of the lifetime of holes (τ_p) on the time (t) of the joint diffusion of platinum and boron.

mechanism had been noticed before for gold diffusion into silicon (Bullis 1966).

Judging by the small obtained values of τ_o ($\tau_0 \approx 3$ ns) and by the extent of the compensation of the base resistivity, the solid solubility limit of platinum ($\sim 10^{16}$ cm^{-3}) was practically achieved in air experiments. The increase of diffusion temperature from 1100° to 1200°C appreciably improves the uniformity and reproducibility of the results and increases the platinum concentration.

Since the content of boron in the source is much higher than that of platinum, platinum does not affect the surface concentration of boron (the surface resistivity change from 0.8 to 6.0 $\Omega/\square$ only can be explained by the change of the polycarborane concentration in the composition). Since the goal of this research (Guk et al. 1986) was to introduce into silicon a quantity of Pt corresponding to its solid solubility, the diffusion of Pt occurred from the polymer composition containing a large amount of platinumorganic compound, namely from an unlimited source (see Section 3.4.1). The subsequent diffusion of phosphorus, performed at 1140°C from 0.5 to 16 hours (also with quenching) did not change the value τ_p in that no appreciable gettering of platinum from the volume of the wafer had occurred.

3.6. MULTILAYER DIFFUSION STRUCTURES

As was shown above, polymer diffusants make it possible to introduce into silicon practically all the necessary impurities. The study of polymer compositions has shown that at a given regime of diffusion and for a given composition of polymer diffusants, the surface concentration value N_s in diffusion layers is determined only by the quantity of impurity applied on the surface of the wafer before doping. This has allowed us to exert precise control over the surface concentration of impurities within several orders of magnitude and to obtain a high degree of uniformity of N_s. Another advantage of the polymer diffusants is also the lack of any surplus amount of doping impurity or any thick layers of the silicate glass when a high surface concentration was obtained. This decreased mechanical stresses and, in certain cases, allowed us to carry out a one-stage diffusion.

Polymer diffusants satisfy the requirements of modern technology concerning the content of contamination by heavy metals and other unwanted elements. The high degree of purification was confirmed by spectral analysis and by the high minority carrier lifetime in the diffusion layers and device bases. The shelf life of polymer diffusants varies from six months to one year, depending on the compositions. The proposed technology makes it possible to obtain the characteristics of the diffusion layers close to those achieved using a much more complicated equipment (Schuegraf 1980).

Below we describe the results of creating various multilayer diffusion silicon devices structures with fairly good and, sometimes, even unique parameters.

3.6.1. Suppressor Diodes

An application of polymer compositions that provides the maximum surface concentration of boron is in the fabrication of fast-switching power devices with a turn-on time of the order of 10^{-11} s. These devices are used to protect electrical circuits. Properties of silicon suppressor diodes, fabricated using the polymer diffusion sources, have been investigated (Zubrilov and Shuman 1987; Zubrilov et al. 1989). The device measurements showed that some diodes did not have any microplasmas, even for diode area exceeding 1 cm^2. Some diodes had a breakdown voltage of the first microplasma very close to the breakdown voltage of the main area of the p-n junction. Such a breakdown can be considered to be *quasi-uniform*. The value of the differential resistance R_d on the linear part of the reverse impulse current-voltage characteristic of diodes served as a characteristic of the breakdown uniformity. As was shown experimentally and theoretically, the main factor determining the value of the differential resistance for pulse longer than 10^{-5} s was the Joule heating of the avalanche multiplication region. A more uniform breakdown is equivalent to a more uniform heat dissipation over the diode area and consequently results in smaller differential resistance. The uniformity of the diffusion layer allowed us to fabricate diode structures with the areas of 1–5 cm^2 with practically uniform breakdown. The avalanche current flowed across the entire area of the p-n junction with the reverse current density $j \geq 10$ A/cm^2. At $j \geq 10$ A/cm^2 the differential resistance R_d remained constant. Also R_d was inversely proportional to the diode area, which indicated a more uniform breakdown than had been reported previously (Grekhov and Sereshkhin 1980). In this reference an increase of the diode area led to the saturation of the R_d value.

A comparison with the results of Kimura and Nishizava (1968) has also confirmed the breakdown uniformity in our structures. These authors investigated the microplasma turn-on process at 20°C. The breakdown voltage of the microplasma was $U_{br} = 68$ V. At the overvoltage $\delta U = 2$–3%, the mean turn-on delay time was 10^{-3}–10^{-4} s. For the same values of U_{br} and δU, our devices had $j \geq 10$ A/cm^2 and the delay time $< 10^{-9}$ s.

3.6.2. High-Voltage Diodes and Thyristors

The dependence of the breakdown voltage for pin-diodes on the thickness of the *i*-region, on its doping level and on the gradient of the impurity concentration in a p-n junction has been calculated (Van Overstraeten and De Man 1970). The joint diffusion of boron and aluminium enabled obtaining a good agreement between the experimental data and the calculated values of the diode breakdown voltages and the thyristor switching voltage over the range from 10 to 4500 V (Guk et al. 1984). To achieve reproducible the forward voltage drop, it is essential to have a reproducible minority carrier lifetime in the bases. Polymer diffusants have a high degree of purity, which is confirmed by the device data (Guk and Shuman 1981). Thus the use of mixed polymer diffusants makes it possible to improve both the uniformity of the current distribution across the area (directly linked to the homogeneity of the boron concentration) and the reproducibility of the forward and reverse current-voltage characteristics in diode structures.

3.6.3. Achieving a High Minority Carrier Lifetime

For main devices, such as silicon solar cells or high-voltage diodes, the minority carrier lifetime must be quite high (if possible, close to that in the pure monocrystal silicon). Techniques that decrease the penetration of impurities and defects from the surface of the wafer into the bulk (Shiraki 1975; Rozgouni et al. 1975) have been developed in order to preserve a high minority lifetime during the high-temperature treatment of Si devices. Methods that allow one to extract uncontrolled impurities from the bulk have also been developed. The latter techniques use a liquid-phase gettering (Silverman and Singleton 1958) as well as gettering with a subsequent long time annealing in the temperature interval, where the solubility of the recombination centers is small and the diffusivity is large, 700°–900°C (Konakova and Shuman 1981).

The use of the polymer diffusion sources enabled us to obtain large τ_p and τ_n in silicon pin-structures even under the conditions of a high-temperature thermal treatment in air. The highest obtained values (τ_p and τ_n = 150 to 300 μs) were close to those for starting Si monocrystals. Thus the polymer diffusants did not introduce any additional recombination centers and allowed us to obtain a high minority carrier lifetime.

3.6.4. Solar Cells

Solar cell fabrication requires the formation of shallow diffusion layers in order to achieve an effective collection of carriers by the p-n junction. Shallow p-n junctions in silicon are usually formed by diffusion from the gaseous phase or by ion implantation. However, the polymer diffusants make it possible to obtain shallow diffusion layers (0.3–1 μm) with a small deviation of the surface concentration on large area wafers (Guk et al. 1995). This technology is both cheaper and simpler. By using diffusants containing boron or phosphorus, solar cells have been fabricated from commercial monocrystalline silicon. Those solar cells were designed to work with solar energy concentrator. They demonstrated characteristics similar to those obtained by using diffusion from gaseous phase (Nasby et al. 1982) or by ion implantation (Kirkpatrick et al. 1977). A large fill factor of the current-voltage characteristics (0.82–0.83) indicated a small spreading resistance of the solar cell front layer. This allowed us to optimize the sheet resistance in order to fit the chosen grid contact. The deviation of characteristics of the solar cells (short circuit current 33–35 mA/mm^2 at AMO, open circuit voltage 590 mV), fabricated on the same wafer 60 mm in diameter, did not exceed $\pm 2\%$ for 4.6×4.6 mm cell dimensions.

3.6.5. Gettering of Undesirable Impurities

A promising material for manufacturing solar cells is polycrystalline silicon edge-feed growth ribbon. The main drawback of this material is its low minority carrier lifetime and, as a consequence, low solar cell efficiency. We have studied the process of high-temperature gettering of polycristalline silicon edge-feed growth ribbon grown by Stepanov's method and used for manufacturing solar cells of large areas (Guk et al. 1990).

A specific feature of the polycrystal silicon is its rather high dislocation density, which can also vary within a wide range from grain to grain. It was evident that this parameter must have a great effect on electrical characteristics of polysilicon, including the minority carrier lifetime. We investigated the efficiency of gettering on the dislocation density in the original material. In order to reveal the dislocations in polysilicon, we used a Secco etchant. Within each grain the dislocation density was nearly constant, but it varied for different grains (from 10^2 to 10^8 cm^{-2}). Diode mesastructures were formed on these polycrystal grains with different dislocation densities.

Graff and Fisher (1980) showed that the efficiency of solar cells reaches an acceptable value ($\sim$ 10%) if the minority carrier lifetime is of the order of 1 μs or higher. The efficiency increases with an increase of the lifetime. This increase continues until the minority carrier lifetime is about 20 μs. Any further increase of that parameter only weakly affects the efficiency of a solar cell. Therefore gettering has to achieve a minority carrier lifetime of at least 1 μs (for polysilicon of a lower quality). To obtain high-efficiency polysilicon solar cells, we did our best to increase the minority carrier lifetime as much as possible by gettering. The efficiency of gettering was estimated by using the value of the electron lifetime τ_n at a high level of injection, measured by the Lax method (Lax and Neustadter 1954). The data on the dislocation density and τ_n values were averaged for a large number of samples.

A polymer source containing phosphorus was used as a getter. High-temperature gettering was carried out for two hours at 1200°C, the surface concentration of phosphorus in the diffusion layer being ~ 10^{21} cm^{-3}. The subsequent low-temperature annealing (700°C) was carried out in order to extract the metal impurities. After the n^+ layer was removed, a shallow n^+p junction was formed on the wafers at 850°C, with a surface concentration of ~ 10^{20} cm^{-3}.

The results showed that high-temperature gettering raises τ_n several times. However, the dislocation density N_{dis} is a parameter that determines whether it is possible to obtain a value $\tau_n \geq 1$ μs (Fig. 3.9). If $N_{dis} > 10^7$ cm^{-2}, τ_n remains small. It follows from Fig. 3.9 that a large gettering effect is achieved for low and medium values of the dislocation density N_{dis}. At the dislocation densities 10^2–10^3 cm^{-2}, τ_n in some structures reached 20 μs. As a result of high-temperature gettering, the excess leakage current in n^+-p junction, caused by the metal precipitations in the layer of a space charge, decreases to 10^{-7} A/cm^2, the diffusion current decreases to

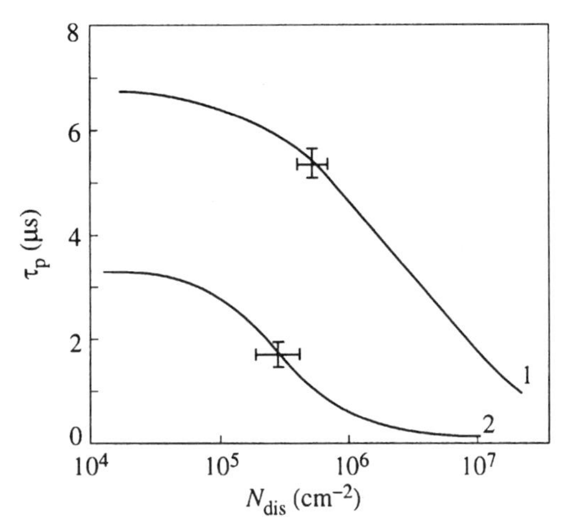

Figure 3.9. The minority charge carrier lifetime (τ_n) dependence on the dislocation density (N_{dis}) in polycrystal silicon: (1) In wafers after a high-temperature gettering and a low-temperature annealing; (2) in standard wafers only after a low-temperature annealing.

10^{-11} A/cm^2, the ideality factor approaches unity, and the value of the open circuit voltage increases to 550–580 mV (at AMO).

3.6.6. Modulator Thyristors

Homogeneity of the diffusion layers is crucial for thyristor fabrication. As has been shown (Akhmedova et al. 1975), in the thyristors working with pulses shorter then 10^{-6} s, the stationary state is not achieved, the turn-on state does not spread over the whole area of the power electrode, and the size of the initial turn-on area is close to the thickness of the structure, 200–300 μm. The width of the initial turn-on area does not depend on the gate current; the only way of increasing it (apart from increasing the emitter perimeter) is to ensure the homogeneity of the turn-on along the whole perimeter. The homogeneity of the turn-on along the emitter-base interface depends on the homogeneity of the doping of all the layers of the structure.

Especially essential is the homogeneity of the doping impurity profile in the thin p-base of the thyristor. Reshetin and Shuman (1980) showed that the absolute delay time of modulator thyristors is determined by the drift time of electrons in the built-in field of the p-base.

For reverse conducting modulator thyristors, the homogeneity of doping in the low ohmic sublayer of the wide base is especially important. Shuman (1980) showed that the homogeneity of the turn-on of such structures depends on the homogeneity of doping in this sublayer. The homogeneity of the turn-on process can be controlled using an image-converter transformer (Kardo-Sysoev et al. 1975a). An improved diffusion layer homogeneity (because of the polymer diffusants) increased the homogeneity of the turn-on process. The modulator thyristors with the power pulse $\sim 10^6$ W and with the current rise time as small as 15–30 ns have been obtained using polymer diffusants. It's worth to note that the turn-on time of such thyristors can be compared with the turn-on time of hydrogen thyratrons with the same power levels (Kardo-Sysoev et al. 1975a, 1975b; Reshetin and Shuman 1980; Shuman 1980).

3.7. POLYMER DIFFUSANTS IN THE TECHNOLOGY OF III-V COMPOUNDS

In the technology of III-V compound devices, the diffusion method of introducing the impurities is not as widely used as in Si technology. Diffusion technology for compounds III-V is much more complicated due to the necessity of maintaining the exact vapor pressure of group V elements (van Gurp et al. 1987).

Until mid 1980s two methods of diffusion were used in III-V compound semiconductor technology: diffusion from a vapor-phase source (Geva and Seidel 1986) and diffusion from a solid source containing the dopant and the group V elements. There are two modifications of the latter method: ampoule diffusion (Matsmoto 1983; Ambreé et al. 1990; van Gurp et al. 1989) and open-tube diffusion (König et al. 1989; Albrecht and Lauterbach 1986).

For every process the use of solid sources requires an exact calculation of the dopant addition and the V group component. Also the sealing of the ampoules has to be guaranteed. This makes technological process nonreproducible from experiment to experiment. A poor reproducibility is typical for both the value of the surface concentration of the doping impurity and the penetration depth. The diffusion from gas-phase sources gives better reproducible results. However, the process is also made complicated by the necessity of maintaining the exact vapor pressure of the group V component.

The most reproducible and convenient method is the diffusion based on using doped spin-on silica films. This method was introduced to III-V compounds technology in the 1980s (Arnold et al. 1984).

Diffusion was carried out in an open tube using a conventional photoresist spinner to coat the surface of III-V compounds with doped spin-on silica films containing a dopant and preventing any evaporation of the group V component.

The main drawback of this method is the necessity of a two-stage process: The first stage consists in forming the oxide film from the emulsion based on the silicic acid ester, and the second stage consists in the diffusion of the impurity from that film into III-V compounds. During the first stage, uncontrollable stresses may arise, leading in some cases to crystal imperfections (Arnold et al.

1984). This problem can be avoided by introducing a thin oxide film on the compound semiconductor surface before spinning the diffusant (Arnold et al. 1984). However, such an operation restricts the applications of this method because this film must be removed for further operations (which is a difficult procedure for planar technology).

Below, using zinc diffusion as an example, we describe a diffusion method that is free of the above drawbacks. This method has been developed at the Ioffe Institute.

3.7.1. Diffusion of Zn in III-V Compounds from Polymer Spin-on Films

Our method of the diffusion from the polymer spin-on films doped by zinc salts includes the following steps:

First, the polymer composition, with dissolved zinc salt, is spun on by centrifuge onto the surface of a III-V semiconductor. Spun in that way, the film does not require any additional treatment before diffusion. Diffusion is carried out in the open tube filled with pure hydrogen or inert gas. Heating and cooling rates are $1°C \cdot s^{-1}$ and $2°C \cdot s^{-1}$, respectively.

The first experiments performed on InP (Belyakov et al. 1992) showed that this method allows one to avoid the decomposition of the III-V compound surface without introducing the V group component into the polymer composition. During the diffusion of Zn, no decomposition was observed for temperatures between 400°C and 656°C ranging from several minutes to several hours. The absence of decomposition was proved by the SIMS profile of phosphorus (Fig. 3.10, curve 1) in InP, observed after the diffusion of Zn at 500°C during one hour. Similar results, showing the absence of evaporation of the V group component, were obtained for GaAs, AlGaAs, InAlAs, InGaAs, InGaAsP, and GaP. For GaAs and GaP no decomposition was observed up to 800°C.

The observed profiles of the zinc distribution in the diffusion layer in InP (Fig. 3.10, curve 2) and in GaAs and InGaAs (Fig. 3.11, curves 1 and 2, respectively) agree very well with the data obtained by other methods (Matsumoto 1983; van Gurp et al. 1989; Albrecht and Lauterbach 1986; König et al. 1989). Such distributions, having

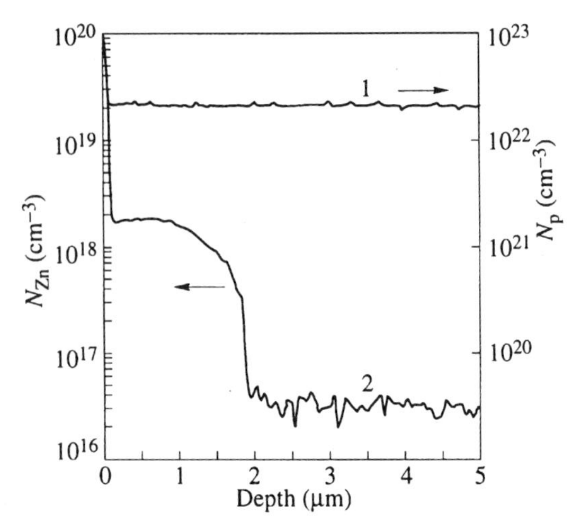

Figure 3.10. SIMS profiles of P (1) and Zn (2) in *n*-InP after the diffusion at 500°C for 60 minutes.

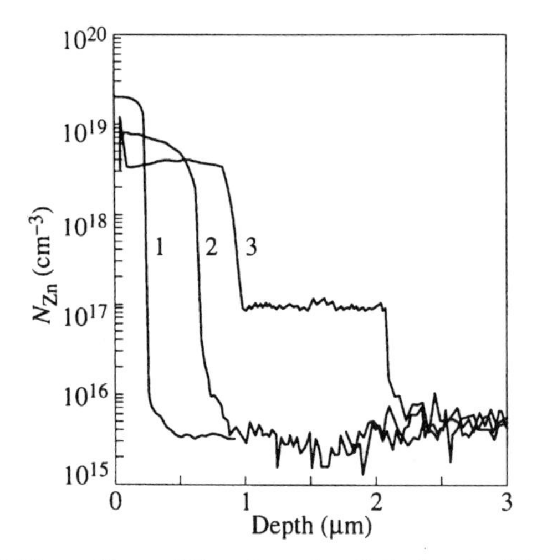

Figure 3.11. SIMS profiles of Zn after the diffusion in (1) GaAs at 650°C, 30 min, (2) $In_{0.47}Ga_{0.53}As$ at 500°C, 30 min, and (3) InP : Fe at 500°C, 30 min.

typical flattened and steep fronts, are usually interpreted within the frames of the interstitial-substitutional mechanism (van Gurp et al. 1989) or within the frame of the kick-out mechanism (Uematsu et al. 1992).

For the semi-insulating InP : Fe, a two-stage profile of the Zn distribution typical for this material is observed (Fig. 3.11, curve 3). This distribution is linked to a high concentration of atoms of iron ($\sim 10^{17}$ cm^{-3}) and to a low concentration of free carriers in this material (less than 10^{13} cm^{-3}). The atoms of iron, as well as the atoms of the group III component, interact with zinc atoms as predicted by the kick-out mechanism. The low concentration of free carriers, which only weakly hinders the penetration of the positively charged intrinsic defects into the semiconductor, facilitates the development of the process.

Figure 3.12 (curves 1 and 2) shows the Zn distribution profiles observed in the case when the concentration of zinc atoms N_{Zn}, introduced into the near-surface region, exceeds the Zn solubility limit L_{Zn} at the diffusion temperature. Even at the initial stage of the diffusion (the stage when the temperature rises from room temperature to diffusion temperature) a fast diffusion of the part of electrically active zinc is observed (Fig. 3.12, curve 1). From the

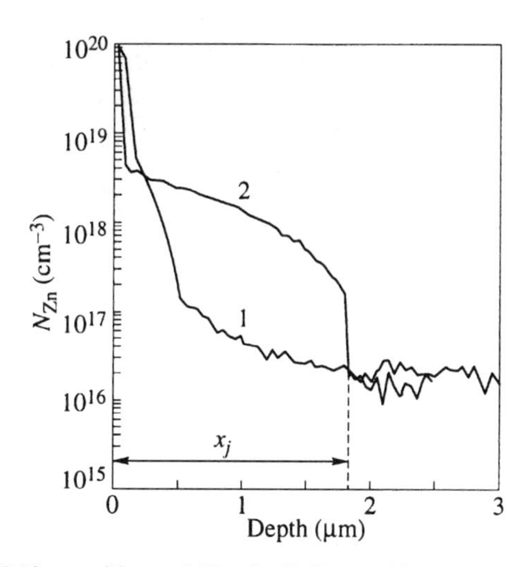

Figure 3.12. SIMS profiles of Zn in InP at $N_{Zn} > L_{Zn}$ after the initial diffusion stage (1) and after the diffusion at 450°C, 30 min (2).

comparison of curve 2, corresponding to a 30-minute diffusion, and curve 1 it becomes clear that at the depth $x_j \geq 1.8$ μm from the surface the concentration of Zn practically does not change during the 30-minute diffusion.

The diffusion of Zn was carried out in n-type InP. The position of the p-n junction, which appeared as a result of diffusion, coincides with the region of the maximal gradient of concentration in curve 2 ($x_j \simeq 1.8$ μm from the surface, where x_j is the depth of the position of the p-n junction) and is the same for both curves.

The distribution of Zn, shown in Fig. 3.12, is typical for comparatively low temperatures of diffusion (450°–500°C for InP). At higher temperature (> 500°C for InP) the shape of the distribution profile after the initial stage is, as a rule, the same as after the complete diffusion process.

In general, a low percent of the electrically active zinc is observed in the near-surface region. For InP the concentration of holes in the near-surface region is $5 \cdot 10^{17}$ cm^{-3} being the concentration of atoms of Zn in that region $\sim 4 \cdot 10^{18}$ cm^{-3}. The concentration of Zn was determined by SIMS, and the concentration of holes—by the *C-V* method or by Raman light scattering (Belousov et al. 1993).

The low impurity activation and the abnormal distribution profiles of Zn (Fig. 3.12) can be linked to supersaturation of the near-surface region with the nonequilibrium intrinsic defects (Tan et al. 1991) at the initial stage of the process with $N_{Zn} > L_{Zn}$.

For the standard process with $N_{Zn} < L_{Zn}$, a high degree of activation of the introduced Zn is typical (practically near to 100%) within the error limit of the control methods being used. We should note that not all diffusion techniques provide such degree of activation. It was reported (Van Gurp et al. 1987) that the impurity activation of the introduced Zn was 25% for InP and 60%–90% for InGaAsP.

During the diffusion of Zn in III-V compounds from the polymer spin-on films the concentration of Zn in the near-surface region could be changed by several orders of magnitude by varying the content of the Zn salts in the polymer spin-on film. The penetration depth of the doping impurity, just as for other methods of diffusion, is determined by the temperature and time regimes and depends on the conductivity type and on the doping level of a III-V compound.

TABLE 3.1.

Material	Diffusion Regime Temperature, °C (30 min)	x_j^2/t (10^{-12} cm$^2 \cdot$ s^{-1})
InP	500	9.4
$In_{0.8}Ga_{0.23}As_{0.39}P_{0.61}$	500	9.4
$In_{0.73}Ga_{0.27}As_{0.63}P_{0.37}$	500	7.3
$In_{0.58}Ga_{0.42}As_{0.9}P_{0.1}$	500	3.6
$In_{0.53}Ga_{0.47}As$	500	2.7
GaAs	650	0.1
GaP	800	0.8
$In_{0.5}Al_{0.5}As$	500	9.4
$Al_xGa_{1-x}As$ ($x = 0.2$–0.6)	650	9.1

For $N_{Zn} \leq L_{Zn}$ the dependence of the p-n junction depth, x_j, on the diffusion time, t_{dif}, has the form $x_j \sim t^{1/2}$ for all investigated III-V compounds: GaAs, InP, $In_xAl_{1-x}As$, $In_xGa_{1-x}As$, $In_xGa_{1-x}As_yP_{1-y}$, and GaP. Table 3.1 gives the values of the effective diffusivity x_j^2/t for these compounds with the electron concentration in the range from $5 \cdot 10^{16}$ to 10^{17} cm^{-3} for $N_{Zn} \leq L_{Zn}$.

It should be noted that the values of the effective diffusivity are close to the values given by other authors (Albrecht and Lauterbach 1986; Glade et al. 1991; Pavesi et al. 1991). The tendency of the diffusivity to increase with an increase of the In content is quite obvious. Compounds that do not contain In have the minimal values of the diffusivity.

The observed tendency leads us to believe that the Zn diffusion process is controlled by In and Ga self-diffusion and goes on in accordance with the kick-out mechanism. It is difficult to make numerical estimations because the In self-diffusion coefficient is not well known and there are practically no data about its concentration and temperature dependencies.

On the whole the behavior of Zn in the investigated compounds is similar to that observed by most investigators using other methods of diffusion.

The ternary compound $Al_xGa_{1-x}As$ is an exception. For AlGaAs, gettering Al by a polymer film from the near-surface region of $Al_xGa_{1-x}As$ to the depth of 0.1–0.2 μm is observed. The change of

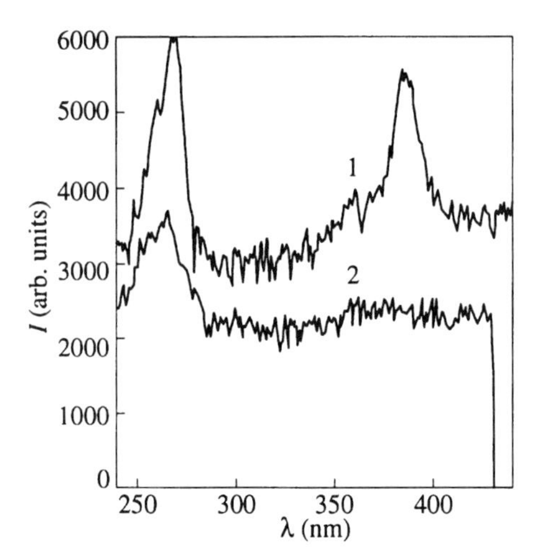

Figure 3.13. Raman scattering spectra for $Al_xGa_{1-x}As$ before (1) and after (2) Zn diffusion at 650°C, 60 min.

the Al content in the near-surface region was observed after the diffusion in the SIMS profiles and in the spectra of the Raman light scattering. Figure 3.13 shows the spectra of AlGaAs before and after diffusion (curves 1 and 2, respectively). The spectra indicated the change of the $Al_xGa_{1-x}As$ composition from $x = 0.45$ to $x = 0.2$ after diffusion. A high solubility of Zn is observed in this region and, what is especially important, a high degree of Zn activation.

The concentration of holes, calculated from the spectrum (Fig. 3.13, curve 2), is 10^{20} cm^{-3}; SIMS in the near-surface region gives the same values of the concentration of Zn. The gettering effect of Al results in homogenization of the compound composition along the surface. We observed the homogenization within the limits of 15–20% after the diffusion at the area 30 × 30 mm with the initial irregularity of the compound composition of about 200%.

3.7.2. Application of Polymer Spin-on Films in the III-V Compound Device Technology

The technology that we developed was used to create the contact p^+-layer in $Al_xGa_{1-x}As/GaAs$ ($x = 0.15–0.55$) light-emitting diode structures. The diffusion from polymer spin-on films provided a high doped layer with the hole concentration of the order of $5 \cdot 10^{19}$–10^{20} cm^{-3} for every investigated compound. The hole concentration in the near-surface layer of these ternary compounds before the diffusion was about 10^{18} cm^{-3}.

An increase in the hole concentration allowed us to reduce the contact resistance and as a result to decrease the voltage drop through the light-emitting diode from 2.2 to 1.5 V at forward bias for the 10 mA current. This method was also used to obtain the p^+-regions of pin-photodiodes based on the InGaAs/InP structures, destined for fiberoptic communication ($\lambda = 1.3$–1.55 μm).

It should be noted that diffusion was carried out through the windows in the mask of SiO_2, which then provided a protection of the periphery of the local p-n junction. The density of the dark current for these p-n junctions was $5 \cdot 10^{-7}$ A/cm^2 at the reverse bias of 5 V. These values are among the best known in literature.

The developed methods of introducing Zn from polymer spin-on films proved to be effective in enabling us to achieve the following:

1. Preserve the surface morphology of the III-V compounds without applying any additional protective cover and without maintaining the vapor pressure of the V group elements. This simplifies the diffusion process and makes the method as simple and reproducible as typical Si technology.
2. Preserve the dielectric properties of the SiO_2 mask and the parameters of the dielectric-semiconductor interface in the process of local diffusion through the windows of a SiO_2 mask.
3. Achieve a high degree of activation of the introduced impurity, close to 100%.
4. Homogenize the surface composition of the $Al_xGa_{1-x}As$ with regard to the content of Al, to reduce the content of Al and obtain a high concentration of holes ($\sim 10^{20}$ cm^{-3}) in the near-surface region.

The method has shown promise for applications in III-V compound device technology for creating heavy doped contact layers and p-n junctions, including local p-n junctions.

3.8. CONCLUSION

A class of new materials has been developed with polymer compositions serving as a source of diffusion for practically every impurity. The use of these materials provides a precisely controlled and homogeneous distribution of the surface concentration of the doping impurity in diffusion layers using a very simple and rather inexpensive technological process. The obtained diffusion layers have characteristics very similar to those obtained by using much more complicated equipment. The polymer diffusants are quite pure and stable materials, widely used both in silicon technology and in III-V compound semiconductor technology. The different semiconductor devices fabricated using these polymer diffusants have shown excellent and, in certain cases, even unique performance. Also a very high level of the device parameter reproducibility has been achieved.

REFERENCES

Akhmedova, M. M., Chashnikov, I. G., Kardo-Sysoev, A. F., et al. (1975). *Sov. Phys. Semicond.* 9, 541–541.

Albrecht, H. and Lauterbach, Ch. (1986). *Jap. J. Appl. Phys. 2, Lett.* 25, L589–L591.

Ambreé, P., Hangleiter, A., Pikuhn, M. H., and Wandel, K. (1990). *Appl. Phys. Lett.* 56, 931–933.

Arnold, N., Schmitt, R., and Heime, K. (1984). *J. Phys. D, Appl. Phys.* 17, 443–474.

Badalov, A. Z. and Shuman, V. B. (1969). *Fizika i Tehnika Poluprov.* 3, 1366–1369 (in Russian).

Badalov, A. Z. and Shuman, V. B. (1970). *Fizika Tverdogo Tela* 12, 2116–2122 (in Russian).

Bellamy, L. J. (1954). In *The Infra-red Spectra of Complex Molecules*, pp. 10–25, 150–180. Wiley, New York.

Bellamy, L. J. (1968). In *Advances in Infrared Group Frequencies*, pp. 130–131. Methuen, Bungay, Suffolk.

Belousov, M. V., Gorelenok, A. T., Gruzdov, V. G., Davydov, V. Yu., and Yakimenko, I. Yu. (1993). *Tech. Phys. Lett.* 19, 33–35.

Belyakov, S. V., Busygina, L. A., Gorelenok, A. T., et al. (1992). *Sov. Tech. Phys. Lett.* 18, 415–416.

Beyer, K. D. (1976). *J. Electrochem. Soc.* 123, 1556–1560.

Beyer, K. D. (1977). *J. Electrochem. Soc.* 124, 630–632.

Borisenko, A. I, Novikov, V. V., Prikhid'ko, I. E., et al. (1972). In *Thin Inorganic Films in Microelectronics*, pp. 30–32. Nauka, Leningrad (in Russian).

Bullis, W. M. (1966). *Sol. St. Electr.* 9, 143–168.

Burger, R. M. and Donovan, R. P. (1967). In *Fundamentals of Silicon Integrated Device Technology: Oxidation, Diffusion and Epitaxy* vol. 1. Prentice-Hall, Englewood Cliffs, NJ.

Carchano, I. (1970). *Sol. St. Electro.* 13.

Carlson, R. O., Hall, R. N., Pell, E. M. (1959). *J. Phys. Chem. Sol.* 8, 81–83.

Conti, M. and Panchieri, A. (1971). *Alta freguenza* 40, 544–547.

El'zov, A. V., Guk, E. G., and Yurre, T. A. (1975). *Sov. Techn. Phys. Lett.* 1, 258–259.

Fujimoto, F. and Komaki, K. (1972). *Appl. Phys. Let.* 20, 248–249.

Geva, M. and Seidel, T. E. (1986). *J. Appl. Phys.* 59, 2408–2415.

Gordon, A. J. and Ford, R. A. (1972). In *The Chemist's Companion: A Handbook of Practical Data, Techniques and References*, pp. 210–230. Wiley, New York.

Graff, K. and Fisher, H. (1979). In *Solar Energy Conversion—Solid State Physics Aspects*, Seraphin, B. D., ed. Topics in Appl. Phys. 31, p. 204, Springer-Verlag, Berlin.

Grekhov, I. V. and Sereshkin, U. N. (1980). In *Avalanche Breakdown of a p-n Junction in Semiconductors*. Energy Leningrad (in Russian).

Guk, E. G., Shuman, V. B., Yurre, T. A., et al. (1977). *Questions of Radioelectronics (TPO)* 1, 55–62 (in Russian).

Guk, E. G. and Shuman, V. B. (1981). *Questions of Radioelectronics (TPO)* 3, 23–24 (in Russian).

Guk, E. G., El'zov, A. V., Luizova, S. F., et al. (1982). *Questions of Radioelectronics (TPO)* 2, 12–14 (in Russian).

Guk, E. G., El'zov, A. V., Yurre, T. A., and Shuman, V. B. (1984). In *Photoresists-Diffusants in Semiconductor Technology*, El'zov, A. V., ed., pp. 88–106. Nauka, Leningrad (in Russian).

Guk, E. G., Shuman, V. B., Yurre, T. A., et al. (1985). *Sov. Techn. Phys. Lett.* 11, 93–94.

Guk, E. G., Yurre, T. A., Shuman, V. B., et al. (1987). *Proc. Conf. on Technology of Power Semiconductor Devices*, Tallinn, pp. 43–46 (in Russian).

Guk, E. G., Shuman, V. B., Tarkhin, D. V., et al. (1990). *Proc. 2nd Int. Conf. on Polycrystalline Semiconductors* 11, pp. 255–258.

Guk, E. G., Shuman, V. B., Shwarz, M. Z., et al. (1995). *Sov. Techn. Phys. Lett.* 21, 61–62.

Irvin, J. C. (1962). *Bell Syst. Techn. J.* 41, 387–410.

Jagi, K., Miyamoto, N., and Nishizawa, N. (1970). *Jap. J. Appl. Phys.* 9, 246–254.

Kaiser, W. and Keck, P. J. (1957). *Appl. Phys.* 28, 882–887.

Kardo-Sysoev, A. F., Reshetin, V. P., Shuman, V. B., et al. (1975a). *Radiotekhn. i Electron.* 20, 1484–1487 (in Russian).

Kardo-Sysoev, A. F., Reshetin, V. P., and Shuman, V. B. (1975b). *Radiotekhn. i Electron.* 20, 1768–1770 (in Russian).

Kimura, C. and Nishizava, J. (1968). *Jap. J. Appl. Phys.* 7, 1453–1463.

Kirkpatrick, A. R., Minnucci, J. A., and Greenwald, A. C. (1977). *IEEE Trans. El. Dev.* ED-24, 429–432.

Koledov, L. A. and Popova, R. B. (1973). *Fizika i Tehnika Poluprovodnikov* 7, 2357–2358 (in Russian).

Konakova, R. V. and Shuman, V. B. (1970). *Electron. Tekhnik.* 5, 66–69 (in Russian).

König, U., Haspeklo, H., Marschall, P., and Kuisl, M. (1989). *J. Appl. Phys.* 65, 548–552.

Lax, B. and Neustagter, S. F. (1954). *J. Appl. Phys.* 25, 1148–1154.

Lisiak, K. P. and Milnes, A. G. (1975). *J. Appl. Phys.* 46, 5229–5234.

Lubashevskaya, A. B., Kataeva, L. A., and Bobrov, L. A. (1968). *Questions of Radioelectronics (OT)* 9, 17–21 (in Russian).

Martin, J., Haas, E., and Raithel, K. (1966). *Sol. St. Electr.* 9, 83–85.

Matsumoto, Y. (1983). *Jap. J. Appl. Phys.* 22, 1699–1704.

Mielke, W. (1975). *J. Electrochem. Soc.* 122, 965–969.

Moreau, W. M. (1988). In *Semiconductor Lithography: Principles, Practices and Materials*, pp. 7–23. Plenum Press, New York.

Nakamoto, K. (1986). In *Infrared and Raman Spectra of Inorganic and Coordination Compounds*, pp. 184–186. Wiley, New York.

Nasby, R. D., Garner, C. M., Sexton, F. M., et al. (1982). *Solar Cells* 6, 49–58.

Pat. USA (1993). N 5094976.

Prikhid'ko, N. E., Borisenko, A. I., Chepik, L. F., et al. (1970). *Questions of Radioelectronics (TPO)* 1, 20–28 (in Russian).

Ramamurthy, V. (1987). ASTM. Special Technical Publication. *Fourth Int. Symp. on Semiconductor Processing*, San Jose, CA, pp. 95–107.

Reshetin, V. P. and Shuman, V. B. (1980). *Radiotekhn. i Electron.* 25, 436–438 (in Russian).

Rozgonui, G. A., Petroff, P. M., and Read, M. H. (1975). *J. Electrochem. Soc.* 122, 1725–1729.

Schuegraf, K. (1980). *Sol. St. Technol.* 23, 87–94.

Shiraki, H. (1975). *Jap. J. Appl. Phys.* 14, 747–752.

Shuman, V. B. (1980). *Radiotekhn. i Electron.* 25, 1560–1562 (in Russian).

Silverman, S. J. and Singleton, J. B. (1958). *J. Electrochem. Soc.* 105, 591–594.

Sprokel, G. I. and Fairfield, J. M. (1965). *J. Electrochem. Soc.* 112, 200–203.

Tan, T. Y., Yu, S., and Gösele, U. (1991). *J. Appl. Phys.* 70, 4823–4826.

Ten, S. T. and Chuan, D. G. S. (1989). *Sol. Energy Mater.* 19, 237–247.

Toropov, N. A. and Barzakovskii, V. P. (1962). In *High Temperature Chemistry Silicates and Other Oxide Systems*, pp. 125–135. Nauka, Moscow (in Russian).

Uematsu, M., Wada, K., and Gösele, U. (1992). *Appl. Phys. A, Sol. Surf.* 55, 301–312.

Unger, B., Schade, U., Hannert, M., et al. (1990). *Proc. SPIE Int. Conf.* 1128, SPIE, pp. 17–24.

Van Gurp, G. J., Boudewijn, R. C., Kempeners, M. N. C., and Tjaden, D. L. A. (1987). *J. Appl. Phys.* 61, 1846–1855.

Van Osterstraeten, R. and De Man, H. (1970). *Sol. St. Elecr.* 13, 583–608.

Welther, W. and Warn, J. R. W. (1962). *J. Chem. Phys.* 37, 292–297.

Wilcox, W. R. and La Chapelle, T. J. (1964). *J. Appl. Phys.* 35, 240–246.

Yoldas, B. E. (1980). *J. Electrochem. Soc.* 127, 2478–2481.

Zubrilov, A. S. and Shuman, V. B. (1987). *Sov. Phys. Techn. Phys.* 32, 1105–1106.

Zubrilov, A. S., Kotin, O. A., and Shuman, V. B. (1989). *Sov. Phys. Semicond.* 23, 380–382.

CHAPTER 4

RARE-EARTH ELEMENTS IN THE TECHNOLOGY OF III-V COMPOUNDS

L. F. ZAKHARENKOV, V. V. KOZLOVSKII, A. T. GORELENOK,
and N. M. SHMIDT

Rare-earth elements (REE) were first applied in semiconductor technology in order to increase the radiation tolerance of silicon solar cells (Mandelkorn 1964). Then Pyshkin et al. (1967) proposed the use of REE in the purification of GaP crystals. Their idea was adopted in the technology of germanium, silicon, and the II-VI semiconductors. Romanenko et al. (1973) and Zakharenkov et al. (1981) showed that the doping of REE makes it possible to reduce the concentration of oxygen in Ge and some other semiconductors. The application of REE in silicon leads to the reduction of the carbon concentration by 1.0–1.5 orders of magnitude (Bagraev et al. 1978).

Intensive investigations of REE for applications in III-V compounds, both for bulk crystals and epitaxial films, started in the 1980s. Two different approaches of REE were explored simultaneously.

The first introduced these impurities in high concentrations into III-V single crystals. In this case the impurities behaved like the

Semiconductor Technology: Processing and Novel Fabrication Techniques,
Edited by M. Levinshtein and M. Shur.
ISBN 0-471-12792-2

isovalent impurities within the crystals' lattices (Zakharenkov et al. 1981; Ennen et al. 1987; Masterov 1990). The presence of REE atoms provided an intracenter luminescence in regions of 1.0 and 1.54 μm. This effect was investigated thoroughly and applied in optoelectronic devices such as lasers and the IR light-emitting diodes (LEDs).

The second approach introduced small concentrations of REE into melt for growing single crystals and epitaxial layers. In this case the REEs acted as the getters of the background impurities and purified the epitaxial layers and single crystals during growth. This effect was first observed in the epitaxial layers of InP grown by the liquid-phase epitaxy (Zakharenkov et al. 1983; Gatsoev et al. 1983), and of InGaAsP (Gatsoev et al. 1983; Bagraev et al. 1984). The concentration of electrons in the layer was reduced from the typical values $10^{17}\,cm^{-3}$ to $10^{13}\,cm^{-3}$. These results were later reproduced by Körber (1986). The patent of Factor (1982) states that the concentration of carriers in InP and InGaAsP has been reduced to $10^{15}\,cm^{-3}$.

Despite seeming simplicity and high efficiency, the purification method of introducing REE into a solution has not received wide acceptance. The reason is that the results greatly depend on the conditions of the experiment and on the concentration of REE. For example, even the material of the crucible used for the bulk crystal growth has an important effect. A wrong choice can nullify the removal of background impurities.

Investigations of REE behavior at the Experimental Physics Department of the St. Petersburg State Technical University and at the Department of the Semiconductor Heterostructures Physics of Ioffe Institute, St. Petersburg, have elucidated the mechanism behind the REE's action in a wide range of concentrations and determined the conditions for reproducible results of purification and doping.

These investigations have led to the development of a growth technology for GaAs and InP crystals, InP epitaxial layers, and InP-based compounds. They have also resulted in the fabrication of optoelectronic and microelectronic devices.

The results of these investigations are described in this chapter.

4.1. REE IN BULK III-V SEMICONDUCTOR CRYSTAL GROWTH TECHNOLOGY

The effect of REE on growing bulk crystals has not been investigated adequately enough. We are aware of only a few publications on this subject. Crystals grown by the method of synthesis-solute diffusion were investigated by Stapor et al. (1986) using InP⟨Yb⟩, Jasiolek and Kalinski (1989) using $Ga_{0.08}In_{0.92}P$⟨Yb⟩, and Jasiolek et al. (1989) using $Ga_{0.65}In_{0.35}As$⟨Yb⟩. Growing the GaSb⟨Gd, Ho⟩ crystals by the horizontal zone melting method was studied by Jasiolek et al. (1986). The GaSb⟨Pr, Yb⟩ and InSb⟨Sm, Yb⟩ crystals were obtained by pulling from the melt (Evgeniev and Kuz'micheva 1990). The data on the crystal properties obtained are incomplete. The data on removing the background impurities from the crystals also have discrepancies.

In this chapter we describe the results of a more complete investigation of REE acting on the electrophysical parameters of bulk crystals of InP and GaAs. We describe various regimes of introducing REE into solutions that can effectively eliminate background impurities. We also discuss the solubility limits of certain REE in the InP and GaAs crystals.

GaAs and InP were chosen as the objects of the investigation because they are widely used for semi-insulating substrates for numerous devices based on III-V compounds. The improvement of the quality of such substrates increases the device yield and improves the device characteristics (Markov et al. 1993a; Fornari 1991).

As a rule, in order to increase the resistivity and carrier mobility μ in the substrates, one must use the initial components of ultrahigh purity (99.99999, or 7N). Even this does not guarantee a uniform distribution of these parameters along the length and across the ingot's cross section. The spread of the parameters (for the ingot grown in different processes) is also considerable—30 percent (Markov et al. 1993b).

The use of REE is one of the most promising ways for the further improvement of the substrate material. Apart from the problem of obtaining semi-insulating substrates, it is of interest to obtain the bulk material that is as pure as possible and whose mobility is as large as possible. Such material is used for the investigation of the interfaces dielectric-semiconductor and metal-semiconductor and for development of diffusion, ion implantation, and other technologies.

As noted above, the investigation of the properties of InP and GaAs with a high concentration of REE is also important for the studies of the intracenter luminescence.

4.1.1. Statement of the Problem of the Crucible

The parameters of dendritic crystals and the InP and GaAs bulk crystals were studied for REE with different electron structures and different chemical activities within a wide interval of REE concentrations from 0.001 to 1 wt.%.

The dendritic crystals were used in our investigations as model objects because the process of their growth had one stage, unlike the process of obtaining bulk crystals. That simplifies the analysis of technological factors and the analysis of their effect on the volume properties of the material.

The InP dendritic crystals were grown from nonstoichiometric melts at temperatures ranging from 1223 to 873 K. The melts cooling down rate was 0.01–0.02 $K \cdot s^{-1}$. Two modifications of the method were used: The first technique utilized BN crucible placed inside of quartz ampoules, vacuum treated to 10^{-5} Torr. The second technique made use of the so-called diffusion seal (Zakharenkov et al. 1986).

The experiments on the dendritic crystal growth from the solutions doped with erbium (Er) and ytterbium (Yb) have shown that the solution interacts with the material of the crucible. This interaction is manifested in the increase of the electron concentration in the InP crystals to $n \sim 3 \times 10^{18}\,cm^{-3}$, compared to the typical values of $n \sim (2\text{–}3) \times 10^{16}\,cm^{-3}$ in crystals grown from the melts not doped with REE.

The emission analysis shows that the content of silicon increases in five to six times when REE are present during growth. The contamination of the growing crystals with silicon, which is an electrically active impurity in GaAs and InP, takes place as a result of the interaction of REE with the material of the quartz crucible:

$$4Yb + 3SiO_2 = 2Yb_2O_3 + 3Si.$$

The use of a crucible made of boron nitride allows one to avoid silicon contamination. However, the interaction of the melt with REE remains when the BN crucible is used.

Bulk InP and GaAs crystals were grown by Czochralski method. InP synthesis was conducted using double-temperature technique in quartz (SiO_2) or BN crucibles (Selin and Antonov 1978), while GaAs was synthesized by directly alloying Ga and As in BN crucibles at nitrogen pressure of 60 atm (the proportion of As/Ga + As being 0.53).

The single crystals were obtained in two stages. First a polycrystalline charge was synthesized. The second stage consisted in extracting a bulk crystal from the melted charge. The fact that the process has two stages hinders the analysis of the effect of REE. However, the interaction with the material of the quartz or the BN crucible was also observed for the bulk crystal growth. Besides, the bulk crystal growth involves the interaction of the liquid B_2O_3 encapsulant with REE in accordance with the reaction:

$$2Yb + B_2O_3 = Yb_2O_3 + 2B.$$

This interaction results in redistribution of the introduced REE between the melt and the B_2O_3 encapsulant and in the formation of electrically nonactive complex with oxygen. The interaction of the melt with the material of the crucible and the encapsulant leads to ambiguous and contradictory data on the influence of REE on the concentration and mobility of carriers in the InP and GaAs crystals (Markov et al. 1993c; Bairamov et al. 1989).

Quartz Crucible with a Protective Coating. A special protective coating was designed for the quartz crucible based on the REE oxides (Bondina et al. 1987a) in order to prevent the interaction of the melt with the crucible and to obtain reproducible results.

An excellent adhesion of the coating to quartz was secured by the formation of an intermediate layer of a complex chemical composition—the REE ortho- and metasilicate (Portnoi and Timofeeva 1986).

The close values of the linear thermal expansion coefficients of the protective coating and quartz provided the protective functions of the coating up to high temperatures (1573–1773 K). Also the process of quartz devitrification slowed down.

The growth of the InP and GaAs crystals performed in the coated crucibles was not affected by the interaction of REE and the melt with the material of the boat. This allowed us to unambiguously determine the dependence of the carrier concentration in the grown crystals on the REE concentration in the melt.

The first results were obtained on the dendritic InP crystals. Doping of the melts with cerium within the concentration range of 10^{-3}–1.0 wt.% resulted in a monotonous growth of the value $N_d - N_a$ from $\sim 7 \times 10^{15}\,\mathrm{cm}^{-3}$ at 10^{-3} wt.% of cerium to $\sim 4 \times 10^{17}\,\mathrm{cm}^{-3}$ at 1 wt.% of cerium (Fig. 4.1).

For the InP crystals, grown under the same conditions but without REE, the typical values of $N_d - N_a$ were $\sim (3–5) \times 10^{16}\,\mathrm{cm}^{-3}$. Thus the introduction of REE into the melt in small concentrations, less than $\sim 10^{-2}$ wt.%, resulted in a considerable reduction of the impurity concentration.

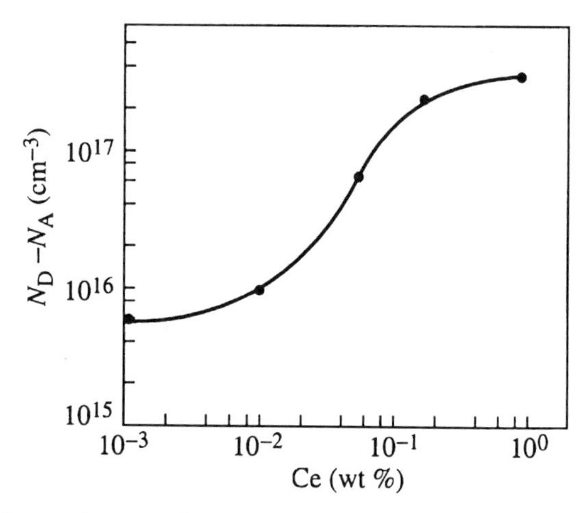

Figure 4.1. Dependence of carrier concentration $n = N_d - N_a$ in InP crystals on the cerium concentration in the melt.

The interaction of REE with the background impurities in the melt was observed when the REE concentration was small. This interaction resulted in the purification of the growing crystal of the background impurities. The mechanism of this effect will be considered in detail in Section 4.2.

The crystals, grown from the melts with a small concentration of REE ($< 10^{-2}$ wt.%) were investigated by electronic paramagnetic resonance method. This measurement showed that the REE ions were not present in the volume of the grown crystal.

Hence either REE were in the bound states or their concentration was below the sensitivity threshold of the method ($\sim 10^{15}\ \text{cm}^{-3}$).

When the concentration of cerium exceeded $\sim 2 \times 10^{-2}$ wt.%, the value $N_d - N_a$ became larger than that in the crystals not doped with REE (Fig. 4.1). The reason is that at larger concentrations of REE a metal-thermic regeneration of the oxides of donor impurities (Si, O, S, Te) increases $N_d - N_a$.

The crystals grown from the melt with larger REE concentrations contain REE in a solid phase. We will consider separately the effect of REE on the properties of crystals for large and small concentrations of REE.

4.1.2. Investigation of the Properties of Bulk Crystals with a High Concentration of REE

Properties of the InP bulk crystals were investigated in detail when europium (Eu) was introduced into the solution. The X-ray crystal analysis showed (Shtel'makh et al. 1990) that when Eu was introduced, the InP lattice remained face-centered, even though the lattice constant a decreased with the increase in the europium content in the melt. For the initial InP, $a = 5.8687 + 0.0001$ Å. When the concentration of europium in the melt was 10^{-1} wt.%, $a = 5.8660 + 0.0004$ Å, with no inclusion of the second phase being detected.

The decrease of the lattice constant with the increase in europium concentration can be explained by a large content of microinclusions of EuO, since this is the only compound of europium with the lattice constant smaller than that of InP (Shtel'makh et al. 1990).

The investigation of magnetic properties of InP⟨Eu⟩ shows that the introduction of Eu changes not only the value but also the sign of the Veiss constant. The Veiss constant Q for the InP⟨Eu⟩ (10^{-2} wt.%) samples is equal to $Q = -6$ K, while for the InP⟨Eu⟩ (10^{-1} wt.%) samples, $Q = +9$ K. Taking into consideration that separate magnetic centers in the crystals under investigation have not been registered by any method, we may conclude that the sign of the Veiss constant manifests the type of magnetic ordering. Thus the InP⟨Eu⟩ (10^{-2} wt.%) samples contain fragments with antiferromagnetic ordering, while the InP⟨Eu⟩ (10^{-1} wt.%) samples contain fragments with ferromagnetic ordering.

The obtained results show that the matrix of indium phosphide comprises europium in the bound form. The inclusions are believed to be EuO and EuS. Also one cannot exclude compounds of europium with background donors of EuTe type.

Similar results were observed when growing single crystals GaSb and InSb in the presence of REE (Evgen'ev and Kuz'micheva 1990). The dependence of the lattice constant and crystal density on the concentration of REE was observed.

For the investigation of the effect of the intracentre luminescence, it is important to know effective distribution coefficients and maximum REE ion concentrations. For InP the maximum concentration of ions, determined by measuring the magnetic susceptibility is $Er^{3+} - 5 \times 10^{18}$ cm^{-3} (Masterov et al. 1989), $Er^{2+} - (2–0.5) \times 10^{20}$ cm^{-3} (Shtel'makh et al. 1990), $Yb^{3+} - 5 \times 10^{17}$ cm^{-3}. Effective REE distribution coefficient according to our estimation have the following values: Yb $- 5 \times 10^{-3}$; Er $- 5 \times 10^{-4}$; Eu $- 10^{-3}$.

4.1.3. Investigation of the Properties of Bulk Crystals at a Low Concentration of REE

Influence of REE on the Mobility of Electrons. As was stated above, the dependence of $N_d - N_a$ on the concentration of REE was qualitatively the same for all lanthanides (see Fig. 4.1). The investigation of the temperature dependence of mobility revealed the dependence of the properties of InP bulk crystals on the position of REE in the lanthanides. We performed comparative investigations of the doping of bulk crystals with cerium, holmium, and ytterbium. These REE are located at the beginning, in the middle, and at the end of the lanthanide row and have different electron structure of the neutral atoms, and different ionic radii.

Figure 4.2 shows the temperature dependencies of mobility of the electrons in InP bulk crystals when the initial melts are doped with different REE. One can see that the doping with cerium decreases the mobility despite the reduction of the value $N_d - N_a$. The doping of Ho and especially of Yb leads to a considerable mobility increase.

The nature of that phenomenon is not yet quite clear. We can, however, suggest the following explanation: The REE atoms are known to be either in interstitials or bound into the neutral complexes (associates) with other defects and residual impurities. The REE atoms from the beginning of the lanthanides, which have larger ionic radii, are less likely to occupy the regular lattice sites in

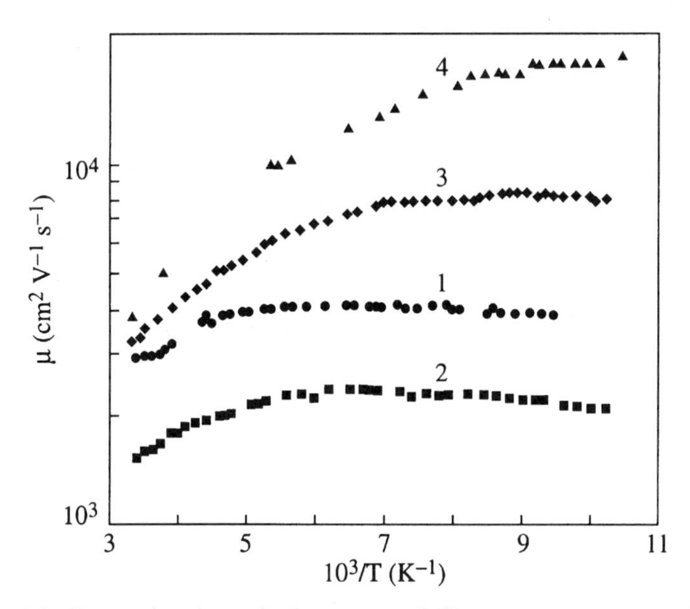

Figure 4.2. Dependencies of electron mobility on temperature in InP crystals doped with rare earth: (1) Undoped $N_d - N_a = 1.5 \times 10^{17}$ cm^{-3}; (2) doped with Ce $N_d - N_a = 1.0 \times 10^{16}$ cm^{-3}; (3) doped with Ho $N_d - N_a = 1.3 \times 10^{16}$ cm^{-3}; (4) doped with Yb $N_d - N_a = 3.5 \times 10^{15}$ cm^{-3}. The concentrations of Ce, Ho, and Yb in the initial melt are the same.

a semiconductor than the "heavy" 4f-elements. For them the interstitial position is more probable. Then the 6s-states of a rare-earth atom can create donor levels in the band gap of the crystal. At the same time those interstitial atoms serve as a sink for other impurity atoms. It is possible that there is a certain equilibrium distribution of the REE atoms in the crystal between these two states. As a result, while doping with the "light" rare-earth atoms such as Ce, one should not expect a great change in the semiconductor electric parameters.

The situation may be different for doping with "heavy" REE elements, whose radii decrease as a result of the lanthanoid compression effect. Such atoms have stable states in the cation sites of the lattice. In this case the REE atom provides three electrons for bonding; that is, it is in a neutral state (charge state $+3$ is deduced in a formal manner from electron configuration of the core). Since the 4f-states are hybridized only slightly with the band states, the ionization cross sections of the 4f-levels are small (unlike the 3d-levels of the transient iron group elements); that is, practically no new charged centers are created in the semiconductor. At the same time the site centers occupied by REE are also sinks for other impurities, since they form neutral complexes.

The best results on "purification" III-V semiconductors can therefore be expected from doping with "heavy" rare-earth elements. However, more detail investigations should be carried out in order to clarify the situation.

Use of REE to Obtain High-Pure InP Bulk Crystals. When the InP single crystals are grown by the Czochralski method using the quartz crucible and the standard technology, the residual $N_d - N_a$ concentration is usually $\sim 3 \times 10^{16}\,\mathrm{cm}^{-3}$, and the mobility of electrons at room temperature is $\mu \sim 4300\,\mathrm{cm}^2\,\mathrm{V}^{-1}\,\mathrm{s}^{-1}$.

The use of the crucible with the protective REE oxides coating described above allows one to considerably improve the parameters of the crystals being grown even without any addition of REE into the melt. Bulk crystals grown in the crucible with a protective coating have $N_d - N_a \sim 3 \times 10^{15}\,\mathrm{cm}^{-3}$ and $\mu \sim 4700\,\mathrm{cm}^2\,\mathrm{V}^{-1}\,\mathrm{s}^{-1}$.

The best results were obtained when a crucible with a projective coating was used and Yb was added to the melt. Then single crystals with $N_d - N_a < 3 \times 10^{15}\,\mathrm{cm}^{-3}$ were obtained with the room temperature mobility $\mu \sim 5000\,\mathrm{cm}^2\,\mathrm{V}^{-1}\,\mathrm{s}^{-1}$.

Obtaining a Semi-insulating InP. The doping of indium phosphide with rare-earth elements was used to obtain a high-quality semi-insulating InP. We developed the technology of obtaining the InP⟨Fe⟩ single crystals with a resistivity of more than $10^7\,\Omega \cdot \mathrm{cm}$, using a simultaneous introduction of ytterbium and iron. The yield of crystals was 60–70 percent, while for the crystals doped with iron alone, it was only 30–40 percent (Bondina et al. 1987b).

Obtaining a Semi-insulating GaAs. At the present time semi-insulating GaAs is often obtained by doping chromium (Deal and Stevenson 1984) by chromium dioxide (Afanas'ev et al. 1982), vanadium (Ulrici et al. 1985), and by titanium (Braudt et al. 1989). GaAs doped with those impurities has a resistivity to $10^7\,\Omega \cdot \mathrm{cm}$, the mobility of electrons not exceeding $4000{-}4500\,\mathrm{cm}^2\,\mathrm{V}^{-1}\,\mathrm{s}^{-1}$ (Fornari 1991). When manufacturing device structures on such substrates, their thermostability appears to be rather low, especially during the ion implantation.

Recently a wide acceptance has been gained by the Czochralski method combined with the synthesis of GaAs by direct alloying of gallium and arsenic, with the excess of As (Markov et al. 1993a). The undoped gallium arsenide grown by this method has a typical resistivity value $\rho > 10^7\,\Omega \cdot \mathrm{cm}$ and a higher electron mobility (in some cases up to $5500\,\mathrm{cm}^2\,\mathrm{V}^{-1}\,\mathrm{s}^{-1}$).

However, as was already mentioned, for such a material, a dispersion of the main parameters (ρ and μ) in the ingot is up to 20–25 percent. From an ingot to an ingot, the dispersion may be higher than 30 percent. That is why we developed a special technique of obtaining the semi-insulating GaAs using REE.

The problem of obtaining a semi-insulating undoped GaAs has two components: (1) lowering the background of the residual impurities; (2) monitoring the type and concentration of native defects. Typically the residual concentration of shallow donors (S, $\mathrm{Si_{Ga}}$) $N_d < 2 \times 10^{15}\,\mathrm{cm}^{-3}$. The residual concentration of acceptors (C, Zn, $\mathrm{Si_{As}}$) is $N_a < 5 \times 10^{15}\,\mathrm{cm}^{-3}$ (Markov et al. 1993b; von Bardeleben and Bourgoin 1990).

The electric properties of pure GaAs are mainly determined by the shallow donors, acceptors, and deep donors EL2. Apart from the above-mentioned levels (the concentrations $N_d + N_a - 10^{15}\ \mathrm{cm}^{-3}$ and $N_{\mathrm{EL2}} - 10^{16}\ \mathrm{cm}^{-3}$), this material contains a two-charge 79/203-meV-deep acceptor with the concentration comparable to the concentration of EL2 deep level (von Bardeleben and Bourgoin 1990).

When the background of the residual shallow impurities is lower than the level of $10^{15}\ \mathrm{cm}^{-3}$, native defects determine the electrical, optical, and paramagnetic properties of the material. All native

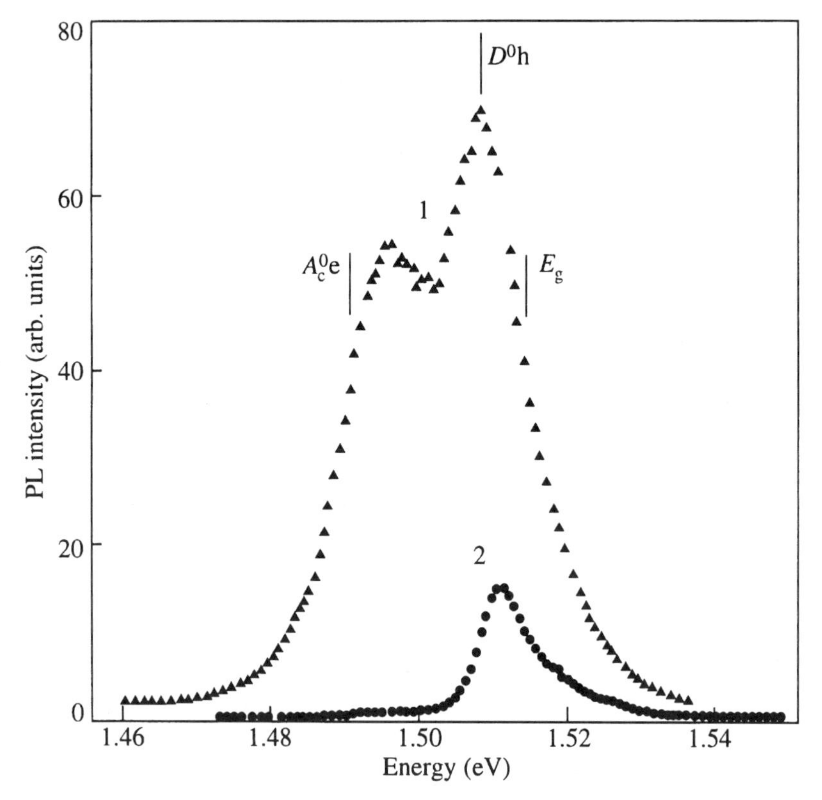

Figure 4.3. Photoluminescence spectra of semi-insulating GaAs single crystals: (1) $\rho = 10^7\ \Omega \cdot \mathrm{cm}$, $\mu = 4500\ \mathrm{cm}^2\ \mathrm{V}^{-1}\ \mathrm{s}^{-1}$ (without REE); (2) $\rho = 10^8\ \Omega \cdot \mathrm{cm}$, $\mu = 6500\ \mathrm{cm}^2\ \mathrm{V}^{-1}\ \mathrm{s}^{-1}$ (with Yb).

defects—vacancies, interstitials, and antisites—in both sublattices are electrically active. The doping of the crystals by REE makes it possible to control somewhat these native defects.

We have investigated the influence of ytterbium on the properties of semi-insulating GaAs. With an optimal concentration of ytterbium in the melt, it was possible to obtain the semi-insulating GaAs with a resistivity of $\rho \sim 3 \times 10^8\ \Omega \cdot \text{cm}$ and the electron mobility at room temperature $\mu \sim 6500\ \text{cm}^2\ \text{V}^{-1}\ \text{s}^{-1}$. The dispersion of ρ and μ occurred both within the ingot and from the ingot to an ingot reduced to 8–12 percent.

The high structural perfection of the semi-insulating GaAs, grown with the addition of ytterbium, is confirmed by the results of an investigation of photoluminescence (PL) spectra (Fig. 4.3). Curve 1 in Figure 4.3 shows a typical PL spectrum for specimens grown without REE ($\rho \sim 10^7\ \Omega \cdot \text{cm}$, $\mu \sim 4500\ \text{cm}^2\ \text{V}^{-1}\ \text{s}^{-1}$). Curve 2 corresponds to the PL of a specimen grown with an optimal concentration of ytterbium ($\rho \sim 10^8\ \Omega \cdot \text{cm}$, $\mu \sim 6500\ \text{cm}^2\ \text{V}^{-1}\ \text{s}^{-1}$), while curve 2 corresponds to the case where all the radiation is free exciton radiation.

The results given in this section demonstrate that REE can be applied with excellent results in the growth technology of high-purity InP single crystals and semi-insulating InP and GaAs.

4.2. REE IN TECHNOLOGY OF InP, InGaAs(P), AND GaP EPILAYERS

The InP, InGaAs(P), and GaP epilayers were grown by method of liquid phase epitaxy (LPE) in the atmosphere of pure hydrogen (< 0.01 ppt) in graphite sliding boats on the substrates of InP⟨Fe⟩, InP : Sn, and GaP : Sn of (100) orientation. REE (Y, Nd, Dy, Ho, Er, Yb, etc.) were placed together with In and Ga into a graphite sliding boat. The purity of REE was 99.9 percent. The concentration of REE in the liquid phase varied within 0.001–0.1 at.%. The epitaxial growth of InP and InGaAs(P) took place at 645°C and GaP at 800°C, the phosphorus supersaturation temperature $\Delta T \sim 5°\text{C}$, the rate of cooling the system being ~ 0.5 deg/min.

4.2.1. Effects of REE Observed in Liquid and Solid Phases

Phenomena Observed in the Liquid Phase. In the InP growth layers the solubility of the InP substrate in the solution of InP–REE increased with the increase of the REE concentration (Gorelenok et al. 1995). This effect is linked to the formation of refractory compounds of P with REE in the liquid phase and with their subsequent slagging. The phosphorus deficit in the solution is compensated by dissolving the InP substrate. Thus equilibrium is established in the system of In-P(In-Ga-As-P)–InP substrate. Since the epilayer's growth is usually performed from the equilibrium or oversaturated solutions, this effect must be taken into consideration when growing multilayer device structures in order to avoid etching a previous layer during the next layer's growth. Figure 4.4 demonstrates the importance of the additional phosphorus supersaturation temperature ΔT of the solution for different compound compositions doped by Ho.

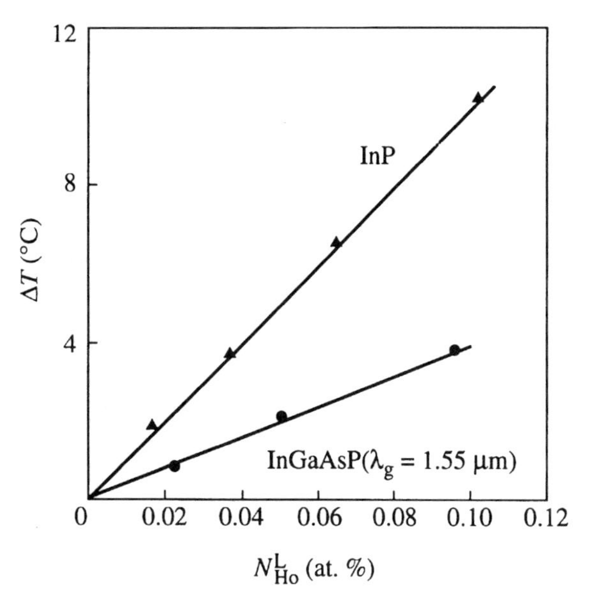

Figure 4.4. Dependence of phosphorus supersaturation temperature ΔT on Ho concentration in liquid phase ($N_{\mathrm{Ho}}^{\mathrm{L}}$) for InP and InGaAsP.

Phenomena Observed in a Solid Phase. The main effect in the solid phase is the "purification" of the growing layer (Fig. 4.5). InP, InGaAsP, and InGaAs epilayers containing only background impurities are n-type with a residual concentration of impurities $N_d - N_a \sim 10^{17}\,\text{cm}^{-3}$. Figure 4.5 shows that with an increase in the REE concentration, the $N_d - N_a$ concentration in the layers initially decreases. As will be shown later in this regime, it is possible to obtain pure, structurally perfect layers with a high mobility and a relatively long lifetime of minority carriers. With further growth of the REE concentration, a compensation takes place. If the REE concentration becomes higher still, *p*-type layers begin to grow. The critical value of the REE concentration has an essential dispersion. This dispersion is related to the chemical activity of REE as well as to the purity level of the initial solution's components, sliding boats, and hydrogen.

Changes in the epilayer parameters that take place with the increase in REE content were investigated by low-temperature edge photoluminescence, by submillimeter spectroscopy, and by SIMS.

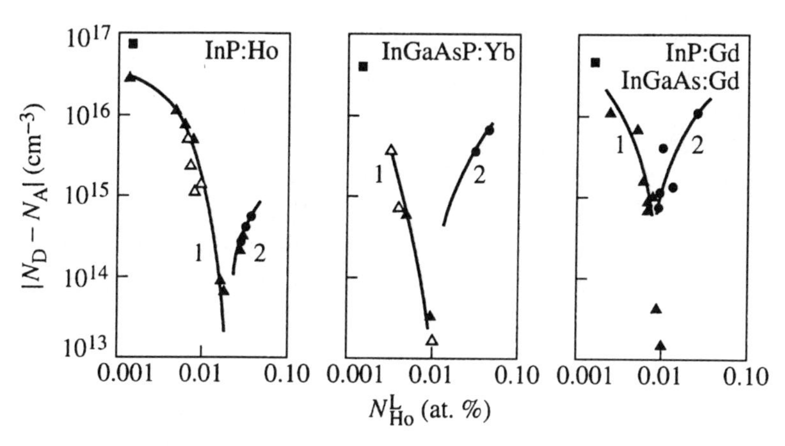

Figure 4.5. Carrier concentration $n = N_d - N_a$ of InP, InGaAsP, and InGaAs epilayers as a function of the REE concentration in liquid phase (N^L_{REE}): (1) n-type; (2) p-type. The ■ indicates the background impurity concentration.

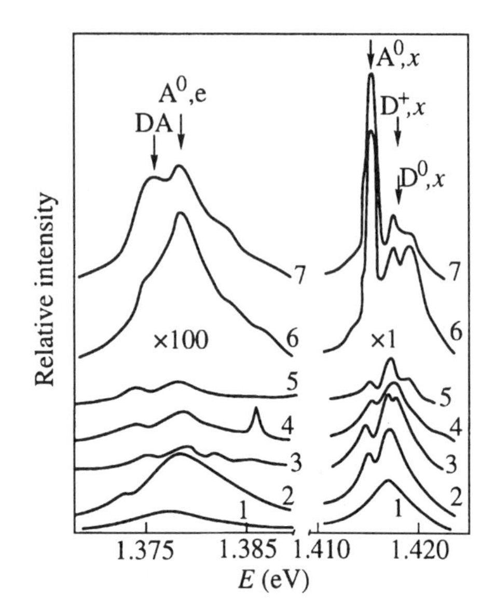

Figure 4.6. Low-temperature edge luminescence of InP epilayers grown with different concentration of gadolinium in liquid phase N^{L}_{Gd} (at.%): (1) 0; (2) 0.001; (3) 0.004; (4) 0.007; (5) 0.009; (6) 0.01; (7) 0.02.

Figure 4.6 shows photoluminescence (PL) spectra for InP with different content of gadolinium (Gd). The edge band of InP ($E \sim 1.415\,\text{eV}$) grown without gadolinium (curve 1) has a half-width ~ 6–$8\,\text{meV}$, determined by the recombination from the states corresponding to free and bound excitons, widened by the chaotic electric fields of the background impurities. The introduction of a relatively small amount of REE (not more than 0.001 at.%) into the solution (curve 2) leads to the narrowing of the line of the neutral exciton–donor complex D^0, X, dominating in the spectrum.

Besides, a line of exciton, bound on the neutral acceptor A^0, X appears in the spectrum. In a relatively weak long-wave part of the spectrum ($E \sim 1.38\,\text{eV}$), the intensity shift to the line A^0, e, corresponding to the recombination of free electrons on shallow acceptors.

For such samples ($N^{L}_{Gd} \sim 0.001$ at.%) an appreciable increase of the mobility is observed. The mobility is two times higher than the mobility for the samples without the addition of REE. The further increase in the REE concentration is also accompanied by substantial changes in the luminescence spectra. With the content of Gd in the solution 0.01 percent (curve 6) the intensity of the line A^0, X

increases abruptly, and the weak step X, corresponding to the free exciton, grows into a line, exceeding D^0, X in intensity. As this takes place, the integral intensity of the entire exciton part of the spectrum and of the line A^0, e increases. Any further increase in the Gd content (0.01–0.03 at.%) (curve 7) causes an appreciable dampening of the lines D^0, X and X while the intensity of lines A^0, e and A^0, X is increasing. The carrier mobility in such samples is much lower than the initial value.

The evolution of the luminescence spectra of InP with the increase in the REE concentration in the solution (Fig. 4.6) illustrates the successive changes of epilayer properties from the epilayer doped only with background impurities ($n \sim 10^{17}\,\mathrm{cm}^{-3}$), with a relatively low mobility to a more pure one ($n \sim 10^{14}\,\mathrm{cm}^{-3}$), and a high mobility and finally to a compensated material, and to a p-type material.

Similar dependencies of the PL spectra on the content of REE in the liquid phase were also observed for InGaAs (Gorelenok et al. 1995). The results of the investigation of samples using a submillimeter high resolution laser magnetospectrometer (1 μeV) (Golubev et al. 1985) are shown in Figure 4.7 for three samples InP with different content of Gd and Yb.

In the spectra of sample 1 with the Gd content in the liquid phase of 0.001 at.% (and the annealing time of one hour), the resonance lines were absent. In the spectra of samples 2 and 3, it is possible to distinguish three lines corresponding to the excitation of three donors and the line of cyclotron resonance (CR).

The right-hand part of Figure 4.7 shows on the magnified narrowest CR line observed in sample 3. The electron cyclotron mass $m^* = (0.0806 \pm 0.0003)m_0$ agrees very well with the value m^* in the pure layers of InP, grown by vapor-phase epitaxy. The effective electron mobility in sample 3 at temperature (4.2 K) reaches the value $\sim 10^6\,\mathrm{cm}^2\,\mathrm{V}^{-1}\,\mathrm{s}^{-1}$, which is comparable to the value of the effective electron mobility in exceptionally pure GaAs layers.

The magnetic field dependences of the shift of donor levels 1s, $3\mathrm{d}_{-1}$, and Landau levels $N = 0$, $N = 1$ are shown in the inset (the calculations use the effective mass approximation (Aldrich and Greene 1979). The calculation was done with the Rydberg effective value $R_y = 7.31$ meV and $m^* = 0.08m_0$. Comparing the intensities

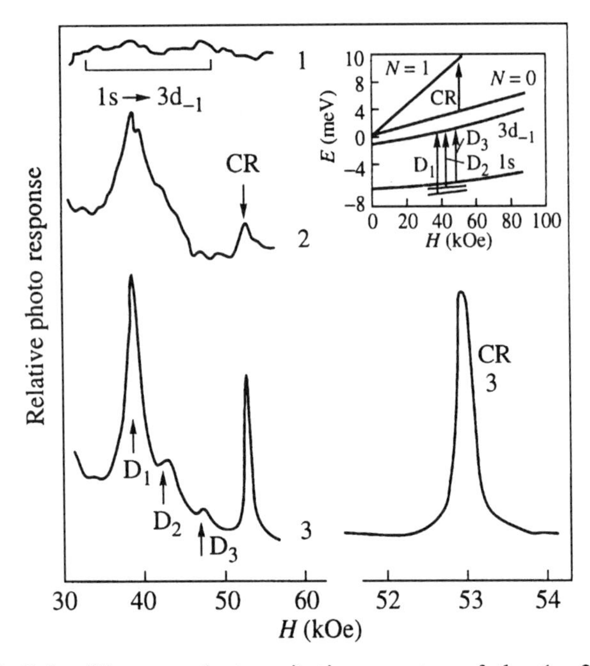

Figure 4.7. Submillimeter photoexcitation spectra of the 1s–3d$_{-1}$ transitions for three samples of InP grown with a different Gd or Yb concentration in solution N^{L} (at.%): (1) Gd, 0.001; (2) Gd, 0.04; (3) Yb, 0.01. The inset shows the magnetic field dependence of the shift in 1s and 3d$_{-1}$ donor levels and the $N = 0$ and $N = 1$ Landau levels. The arrows show the transitions detected, 4.2 K.

of lines of the donors in photoexcitation spectra 1s → 2_{p+1} and 1s → 3d$_{-1}$, one can conclude that donor D_1 is introduced by sulfur and D_2 (or D_3) by silicon.

Epilayers grown with a low concentration of REE in liquid phase $N^{\mathrm{L}}_{\mathrm{REE}} \leq 0.005$ percent have a good morphology and a high uniformity of carrier concentration over the area and as a function of the layer depth. In the range of $N^{\mathrm{L}}_{\mathrm{REE}}$ (0–0.005 at.%), no changes in the lattice parameter a of the InP have been observed (Gorelenok et al. 1995). The Raman spectra of the InP pure epilayers yield the same results (Bairamov et al. 1989). This fact seems to indicate that either there are no REE in the grown epilayer or their concentration does not exceed $10^{13}\,\mathrm{cm}^{-3}$.

The increase in the REE concentration in the liquid phase to the value of $N^{L}_{REE} \gg 0.005$ percent is followed by the appearance of the second phase inclusions (Nakagome 1987; Gorelenok et al. 1995). These inclusions are the compounds of P, O, and elements of group I with REE. An abrupt increase in the density of the inclusions (to $\sim 10^4\,cm^{-2}$) is observed at $N^{L}_{REE} > 0.01$ at.%. The sizes of inclusions vary between 5 and 200 μm, depending on the values of N^{L}_{REE}.

The quantitative estimation of the content of atoms Yb in the InP epilayers, by SIMS is shown in Figure 4.8. The figure indicates that when the Yb concentration in liquid phase is small, the concentration profile is uniform (curve 1). At $N^{L}_{Yb} = 0.01$ at.% (curve 2) there is a concentration peak of the Yb content near the layer-substrate interface. This peak is apparently linked to the inclusion of the second phase.

Figure 4.8 shows that the concentration of the Yb in the uniform samples is very close to the SIMS sensitivity limit: $(2-4) \times 10^{15}\,cm^{-3}$. The estimations of the Yb distribution coefficient in InP according to these data show that it does not exceed 10^{-4}, in contrast to the values $(4-8) \times 10^{-2}$, obtained for Dy, Gd, and Sm in GaAs (Romanenko and Kheifets 1973).

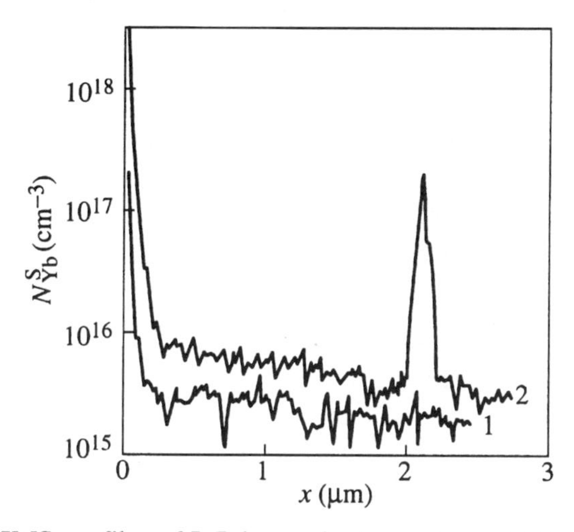

Figure 4.8. SIMS profiles of InP layers for two values of Yb concentration in the liquid phase: (1) 0.001 at.%; (2) 0.01 at.%.

Using REE doping, the n-type InP epilayers were obtained with concentration $n = 10^{14}\ \mathrm{cm}^{-3}$ and mobility $\mu = (0.7–1) \times 10^5\ \mathrm{cm}^2\ \mathrm{V}^{-1}\ \mathrm{s}^{-1}$ at 77 K (Bagraev et al. 1984). The minority-carrier lifetime for the best layers was $\tau \sim 10\ \mu\mathrm{s}$ compared to $\tau \sim 300$ ns for the layers grown under similar conditions without the REE doping (Gorelenok et al. 1987).

The n-type InGaAs epilayers with concentration $n = 10^{14}\ \mathrm{cm}^{-3}$ and mobility $\mu = 15400\ \mathrm{cm}^2\ \mathrm{V}^{-1}\ \mathrm{s}^{-1}$ at 300 K and $\mu = 10^5\ \mathrm{cm}^2\ \mathrm{V}^{-1}\ \mathrm{s}^{-1}$ at 77 K were grown using the REE doping. Layers with the concentration $n = (1–2) \times 10^{17}\ \mathrm{cm}^{-3}$ had the mobility $\mu = 7000\ \mathrm{cm}^2\ \mathrm{V}^{-1}\ \mathrm{s}^{-1}$ at 300 K. The high quality of the obtained InGaAs layers can be illustrated by the dependence of the drift velocity on the electric field $V_d(E)$ in strong electric fields (Fig. 4.9) (Galvanauskas et al. 1988; Balinas et al. 1990). Note that the $V_d(E)$ dependences agree very well with the Monte Carlo calculations.

The InGaAsP layers of different composition with small concentrations $n = N_d - N_a = (10^{14}–10^{15})\ \mathrm{cm}^{-3}$ and sufficiently high electron mobility in the low field were also obtained using REE doping. The field dependences of the electron drift velocity for these layers are shown in Figure 4.10. These dependences also agree quite well with the results of the Monte Carlo calculations.

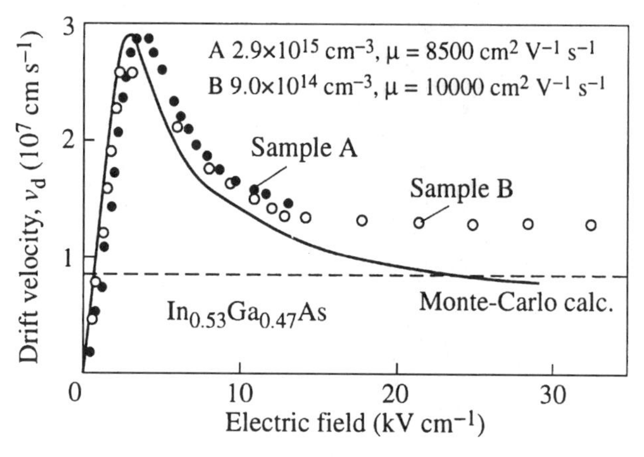

Figure 4.9. Field dependence of electron drift velocity for two samples of $In_{0.53}Ga_{0.47}As$, 300 K.

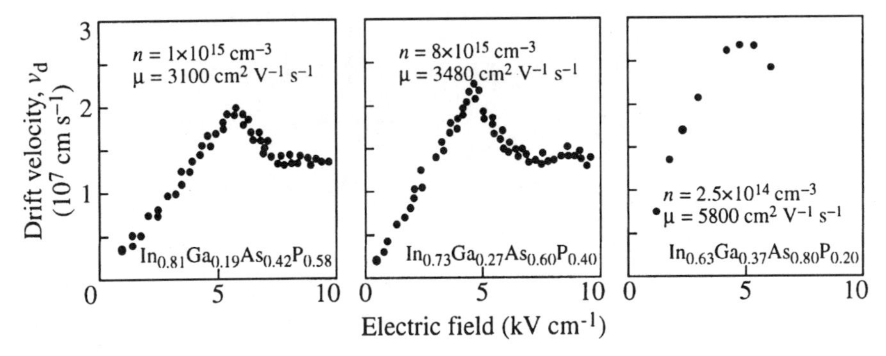

Figure 4.10. Field dependences of the electron drift velocity for InGaAsP alloys of different compositions, 300 K.

4.2.2. Mechanism of LPE Layer Purification by Rare-Earth Elements

The experiments with simultaneous doping of the growth compound by REE and by the donors of IV (Si, Ge, Sn) and VI (S, Se, Te) groups and also by the acceptors of the II (Mg, Cd) group (Gorelenok et al. 1988) were performed in order to study the mechanism of the InP layer purification by REE.

The donors were introduced into the liquid phase in such a quantity that the concentration of electrons in the grown layers would be $\sim 5 \times 10^{18}$ cm^{-3}, namely by more than an order higher than the usual level of the background concentration. Then, without changing the donor concentration, various quantities of Dy were added to the solution. The results of these experiments are presented in Figure 4.11.

The figure shows that the electron (and donor) concentration in the layers doped by the elements of group IV (Ge, Si, Sn) changed very little with the addition of Dy into the solution. At the same time, for InP doped with the VI group donors, the addition of Dy results in a substantial decrease in the donor concentration (up to an order of magnitude or more).

The smaller is the atomic weight of the donor and the higher is its chemical activity, the greater is that decrease. The experimental data given here prove that REE interact with the elements of group VI in the solution (as well as with oxygen). This gives rise to the

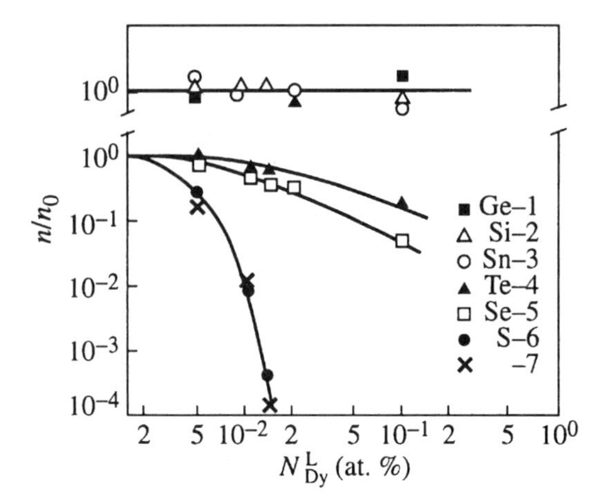

Figure 4.11. Normalized electron concentration in InP layers doped with donors of group IV (1—Ge; 2—Si; 3—Sn) and of group VI (4—Te; 5—Se; 6—S) as a function of dysprosium concentration in solution (7—unintentionally doped sample).

formation of high-temperature chalcogenide compounds (Karpov et al. 1984). These compounds fall out as slag in the liquid phase and do not enter the solid phase.

It should be noted that the dependencies n/n_0 on N^L_{Dy} for the samples doped with S and for the undoped samples practically coincide (curves 6 and 7 in Fig. 4.11). This means that the main background impurity in the InP epilayers may be sulfur.

Simultaneous doping with Yb and with an acceptor impurity leads to the opposite results: The addition of REE into the solution does not reduce the carrier concentration but rather increases it (see Table 4.1). The increase in the hole concentration is caused by deoxidization of the acceptor impurity by REE in the solution.

4.2.3. Obtaining Epilayers with Reproducible Parameters

The level of contamination by background impurities may vary within a very wide range. For the same level of solution doping, the carrier concentration in the growing epilayer may vary from process to process by an order.

TABLE 4.1. Simultaneous doping by Yb + Mg and Yb + Cd.

	Doping (at.% T = 300 K)			
$In_{1-x}Ga_xAs_{1-y}P_y$	Acceptor	Yb	p (10^{-19} cm^{-3})	μ (cm^2 V^{-1} s^{-1})
$x = 0.27$, $y = 0.37$	Mg, 0.05	0	3.5	185
	Mg, 0.05	0.01	4.1	230
$x = 0.46$, $y = 0.08$	Mg, 0.05	0	1.8	100
	Mg, 0.05	0.01	6.5	110
$x = 0.47$, $y = 0.00$	Mg, 0.05	0	2.0	140
	Mg, 0.05	0.01	4.5	150
$x = 0.46$, $y = 0.08$	Cd, 1.0	0	4.1×10^{-2}	76
	Cd, 2.0	0.08	3.1×10^{-1}	47

To obtain the epilayers with the given reproducible concentration, we used simultaneous doping by REE + Sn (in the liquid phase, Sn practically does not interact with REE). The technique works when the amount of REE in the liquid phase satisfies the condition $N_a > N_d$. Given this condition, the electron concentration in the epilayers can be controlled by varying the Sn concentration.

Figure 4.12 illustrates how the concentration of electrons can be controlled in InGaAsP layers in the range 2×10^{14} cm^{-3} to

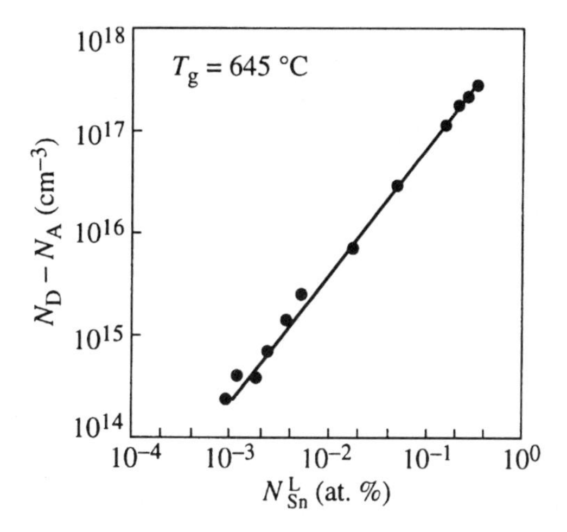

Figure 4.12. Carrier-concentration dependence of InGaAsP epilayers as a function of Sn concentration in solution (N^L_{Dy} = 0.006 at.%).

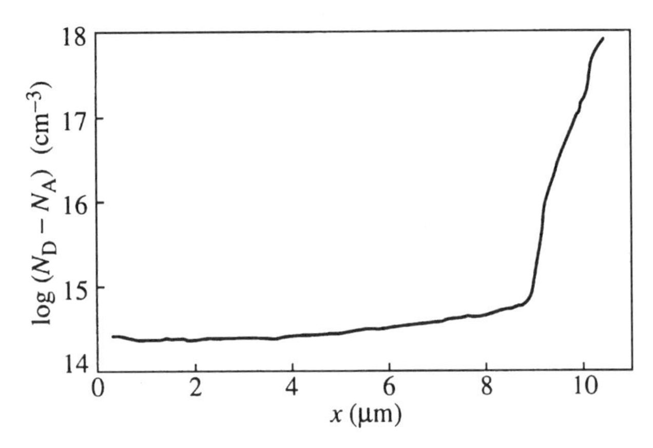

Figure 4.13. Distribution of electrons along the InGaAs epilayer grown from a solution simultaneously doped by Dy (N^L_{Dy} = 0.006 at.%) and Sn (N^L_{Sn} = 0.001 at.%).

$3 \times 10^{17}\,cm^{-3}$, with N^L_{REE} = 0.006 at.% and the Sn concentration changing from 0.001 to 0.4 at.%.

Figure 4.13 shows a typical distribution of the electron concentration along the InGaAs epilayer obtained by the simultaneous doping with Dy + Sn. One can see that this concentration profile is sufficiently uniform.

Simultaneous doping is also effective for obtaining reproducible results with high acceptor doping such as Mg and Cd. Such simultaneous doping made it possible to obtain a high degree of acceptor activation.

Table 4.1 gives carrier concentrations in epilayers doped with Mg and Cd. All other conditions being the same, the simultaneous doping yields a higher concentration of holes in the doped layer. This is evidence that in the liquid phase REE hardly interacts with Mg and Cd. As was already noted, the increase in the hole concentration in the layers is caused by the deoxidization of the acceptors by REE, which in turn causes an increase in the acceptor concentration in the liquid phase and, accordingly, an increase in the hole concentration. This approach may be used for forming contact layers in device structures.

4.3. REE IN DEVICE TECHNOLOGY

The technique of obtaining epitaxial layers using REE was also applied in producing GaP LEDs, Gunn diodes, InP, and InGaAsP photodetectors, and field-effect transistors of different types.

GaP Light-Emitting Diodes. One of the challenges in the LED technology is the fabrication of LEDs emitting "pure green" light. Such LEDs at room temperature (300 K) must emit light with the wavelength $\lambda = 555$ nm ($h\nu_0 = 2.23$ eV) corresponding to the peak sensitivity of the human eye. At the same time such LEDs must emit as few quanta of other energies as possible.

At room temperature the GaP band gap is quite close to the desired quantum energy $h\nu$. However, it is still impossible to obtain the pure green light from GaP LEDs produced using traditional technology. As a rule, in addition to the band-to-band radiation, one observes the radiation spectrum linked to the presence of the background impurities of groups II and VI in the material.

Such lines can be observed quite clearly in the low-temperature luminescence spectra when the spectral resolution is high.

Figure 4.14*A* shows a typical PL spectrum of a low-doped GaP layer of *n*-type ($n = 4 \times 10^{16}\,\mathrm{cm}^{-3}$) at temperature $T = 2$ K. The layer was obtained by a traditional LPE process. The *n*-type layers with such a concentration are usually used for an active region of GaP LEDs. The band gap increases at lower temperatures, and at $T = 2$ K the band-to-band radiation corresponds to the energy $h\nu = 2.32$ eV (peak 1, Fig. 4.14*A*). In addition to peak 1, there are also undesirable impurity peaks 2 and 3 with a considerable intensity. These peaks correspond to the recombination of donor-acceptor pairs linked to the background impurities of groups II and VI.

The doping of the liquid phase with yttrium during the growth of the active region allows one to lower the donor background concentration. The purification effect is illustrated by the photoluminescence spectrum of GaP grown with the introduction of yttrium into the liquid phase ($N_Y^L = 0.0055$ at.%) (Fig. 4.14*B*). The doping level of the epilayer is approximately the same as for the sample with the spectrum shown in Figure 4.14*A*. As can be seen in the figure, the

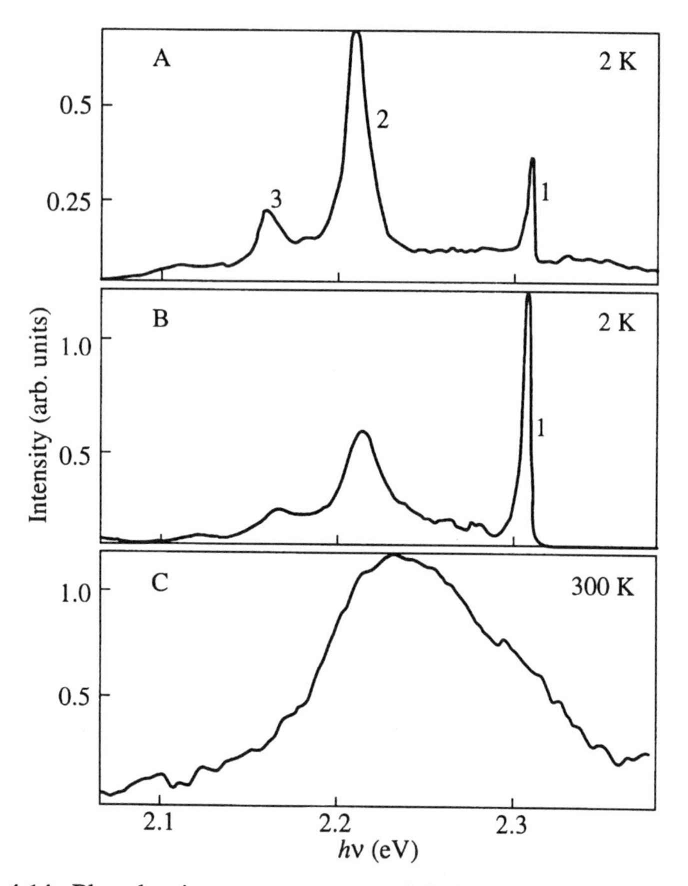

Figure 4.14. Photoluminescence spectra of GaP epitaxial layers ($n \sim 4 \times 10^{16}$ cm^{-3}): (A) $T = 2$ K epilayer grown from a solution undoped by REE. (B) $T = 2$ K epilayer grown from a solution doped by Y ($N_{\mathrm{Y}}^{\mathrm{L}} = 0.005$ at.%). (C) $T = 300$ K. The same epilayer as sample shown in panel B.

spectrum of the sample, grown with the addition of yttrium, is dominated by the band-to-band radiation ($h\nu = 2.307$ eV at $T = 2$ K). The impurity lines are suppressed to a considerable extent. At $T = 300$ K practically only band-to-band pure green luminescence is observed in such samples, with the half-width of ~ 80 meV (Fig. 4.14C). At $T = 300$ K the energy of the peak $h\nu_0 = 2.24$ eV corresponds to the band-to-band radiation.

However, to obtain purely green luminescence, it is not enough to suppress the impurity related radiation in the active (base) LED region. It is also necessary to suppress the impurity luminescence from the LED p^+-emitter. The formation of the LED p^+-region by traditional methods (e.g., diffusion or epitaxy) is accompanied by the introduction of background impurities of groups VI and II, causing appearance of the intensive impurity lines similar to those shown in Figure 4.14*A*.

Therefore, in order to provide the pure green luminescence in GaP LEDs, simultaneous doping in the liquid phase with manganese and yttrium was used during the growth of the p^+ region.

Figure 4.15 shows a typical photoluminescence spectrum of the *p*-type GaP layer with a hole concentration $\sim 10^{18}\,\text{cm}^{-3}$ at 300 K obtained by this method. The luminescence spectrum practically consists of one band with $h\nu = 2.234$ eV at 300 K. Electroluminescence spectra of LEDs grown using this technology have only one pure green band. Thus the use of REE in liquid-phase epitaxy makes it possible to fabricate pure green ($\lambda = 555$ mm) GaP LEDs (Gorelenok and Shpakov 1996).

InGaAsP Gunn Diodes. The technique for obtaining pure semiconductor layers was also used for the fabrication of the Gunn diodes (GD) on the n^+-InP substrates of (100)-orientation with

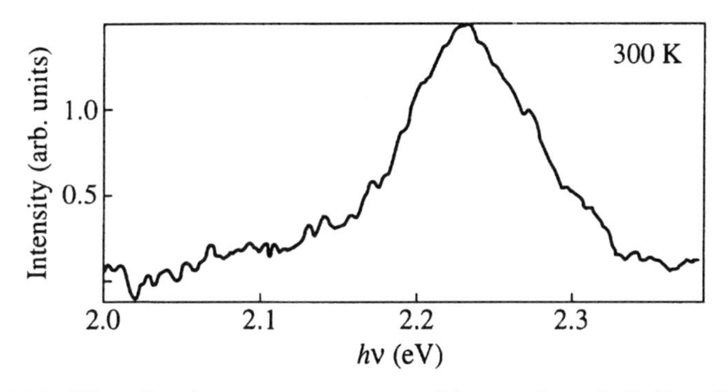

Figure 4.15. Photoluminescence spectra of heavy doped GaP epilayer of *p*-type ($p = \sim 10^{18}\,\text{cm}^{-3}$) grown from the solution simultaneously doped by Mg and Y ($N_Y^L = 0.005$ at.%).

$n = 10^{18}\,cm^{-3}$. The electron concentration in the InGaAs active region was $10^{15} - 10^{16}\,cm^{-3}$ with the electron mobility $\mu = 9000 - 10{,}000\,cm^2\ V^{-1}\ s^{-1}$. The diodes had a honeycomb structure (Borisov et al. 1992) with the contact diameter of 12 μm. This made it possible to perform ac measurements without any additional heat sinking. The Gunn diode efficiency was about 2.5 percent for the 8-mm wavelength (Borisov et al. 1992).

These InGaAs Gunn diodes have a higher frequency than traditional GaAs and InP diodes for the same thickness, since the peak electron drift velocity in n-InGaAs is higher than in GaAs and InP (Balinas et al. 1990).

Photodetectors. The layers of InP and InGaAs obtained using REE doping were used for different photodetectors: phototransistors, the *pin*-diodes, field effect transistors, and bipolar phototransistors.

Low doped layers of InP and InGaAs ($n = N_d - N_a = \sim 10^{13} - 10^{15}\,cm^{-3}$) were used for *photoresistors*. These layers were grown on semi-insulating substrates InP : Fe ($\rho \sim 10^7\ \Omega \cdot cm$) (Alferov et al. 1983; Gorelenok et al. 1985). The dark resistance of the photoresistors was $10^6 - 10^8\ \Omega$ for InP and $10^3 - 10^6\ \Omega$ for InGaAs.

At low electric fields the current-voltage characteristics were linear. With an increase in the field to $E \sim 2 \times 10^3\ V\ cm^{-1}$ for

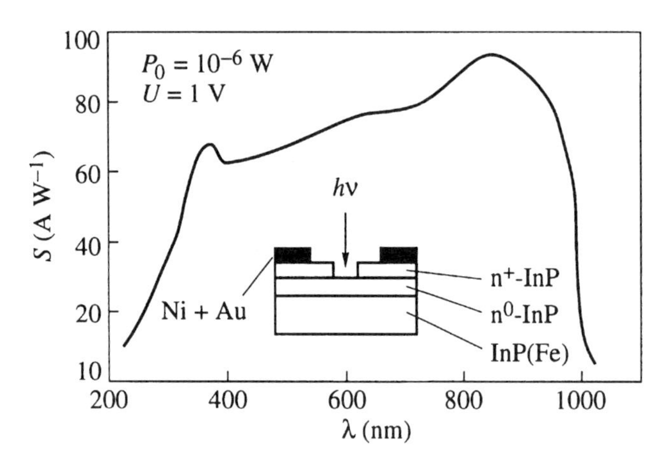

Figure 4.16. Spectral response of InP photoresistors: Bias $U = 1$ V; light power $P_0 = 10$ W.

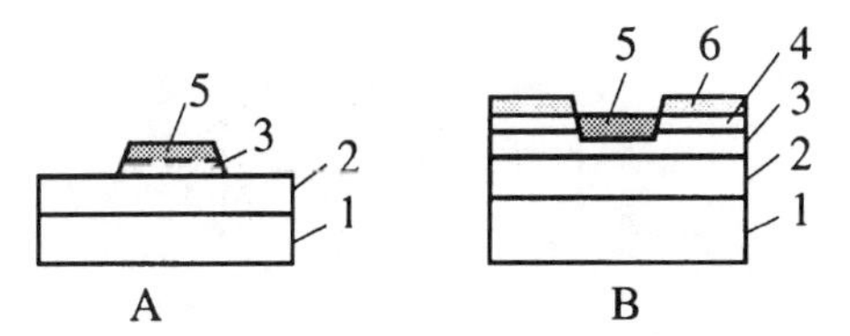

Figure 4.17. Schematic diagram of the *pin*-photodiodes: (*A*) Mesa *pin*-photodiode; (*B*) planar photodiode. (1) n^+-InP : Sn substrate; (2) undoped InP (*n*-type) buffer layer; (3) narrow-gap *n*-type InGaAs (InGaAsP) at $E_g = 0.73$ eV; (4) *n*-type InGaAsP at $E_g = 1.0$ eV; (5) *p*-type InGaAs (InGaAsP); (6) SiO_2.

InGaAs and $E = 10^4$ V cm^{-1} for InP, a deviation from Ohm's law was observed due to the transferred electron effect.

Photoresistors based on InP and InGaAs have spectral sensitivity in a very broad range that extends into the ultraviolet region. Even for a wavelength of $\lambda = 300$ nm, the sensitivity reached 40 A W^{-1} (Fig. 4.16) (Alferov et al. 1983). The long-wave boundary is determined by the band gap of the material. InGaAs photoresistors with the length of active layer of 5 μm had a response time of 10^{-10} s and a gain of 8–10.

The technique of obtaining pure layers using REE was also attempted in the fabrication of mesa and planar photodiodes (Andreev et al. 1985a; 1985b). Figure 4.17 shows both types. In the case of the mesa diode (Fig. 4.17*A*), polyimide passivation was used to stabilize the dark currents (Andreev et al. 1985b). The spectral sensitivity in the region 1.3–1.5 μm was 0.5–0.7 A W^{-1} without any antireflection coating. The dark current density was approximately 10^{-7}A cm^{-2} with a reverse bias voltage of 1–5 V.

The *pin*-photodiodes with diameters of 50–100 μm were illuminated by 25-ps laser pulses with a wavelength of 1.3 μm. The photoresponse rise and decay times were $\sim$ 50 ps (Volkov et al. 1987).

Vertical photofield-effect phototransistors were fabricated using the same technique of REE doping. These devices utilize a modulation of the built-in potential by absorbed light (Bogdanovich et al. 1985; Gorelenok et al. 1995).

Figure 4.18 *A* gives a schematic diagram of the field phototransistor with a buried gate grown on the n^+-InP : Sn substrate. First an

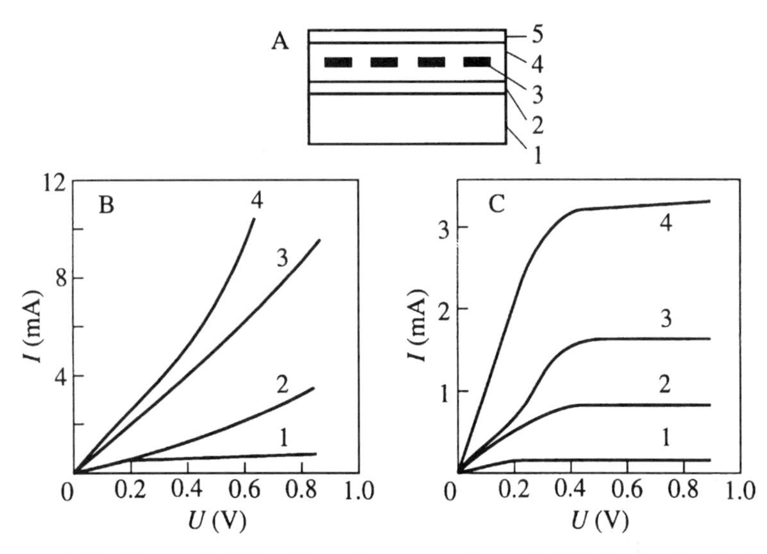

Figure 4.18. Vertical field-effect phototransistor: (*A*) Schematic of photo-FET with buried gate. (1) n^+-InP substrate; (2) InP buffer layer; (3) p^+-InGaAs (buried gate); (4) n^0-InGaAs; (5) n^+-InGaAs. Current-voltage characteristics: (*B*) triode type; (*C*) pentode type. Light power (*W*): (1) 0; (2) 5×10^{-6}; (3) 3×10^{-4}; (4) 2×10^{-2}.

undoped buffer layer of InP with $n = 5 \times 10^{16}\ \mathrm{cm}^{-3}$, 3 μm thick, was grown. Then an active layer of InGaAs(InP) with $n = (1-10) \times 10^{14}\ \mathrm{cm}^{-3}$, 4–6 μm thick, was deposited, and a buried one-micron gate was fabricated by selective diffusion of zinc in the 5×5 or 10×10-μm SiO_2 windows with the 5-μm spacing. After removing SiO_2 the structure was covered by a pure layer of n^0-InGaAs (InP) ~ 3–4 μm thick and by a contact layer of n^+-InGaAs(InP) ~ 1 μm thick. The contacts to the structure were made by vacuum metal deposition.

Figures 4.18*B, C* show the output current-voltage characteristics, which were either of a triode (Fig. 4.18*B*) or pentode type (Fig. 4.18*C*) depending on the level of doping in the n^0-InGaAs(InP) layer and on the transistor geometry. Triode's current-voltage characteristics were usually observed when the total sum of the space-charge regions of the p^+–n^0-junctions exceeded the geometric width of the channel. When this sum was less than the width of the channel, the characteristics were of the pentode type.

For the triode type structures, the gain first increased with the increase of the light intensity; then it reached a maximum and decreased. For the pentode type device, the gain practically remained constant in a wide range of the incident light intensity. The pulse studies showed that the rise time depends on the incident light intensity, on the geometry of structures, and on the applied voltage. At high intensities ($\sim 10^{-2}$ W) the rise time and decay time reached ~ 10 ns. At the light intensities less than 10^{-4} W, the rise time increased to 100–200 ns. The spectral sensitivity range was approximately 1.0–1.6 μm. The steady-state gain reached 100 with the light power smaller than 10^{-4} W and bias voltage 1.5 V.

Bipolar NpN-Phototransistors. Simultaneous doping of Cd + REE was used for the fabrication of the *NpN*-bipolar phototransistors with a highly doped thin base (Gorelenok et al. 1984) (Fig. 4.19*A*). The transistor base consisted of the *p*-InGaAsP (E_g = 0.8 eV) with a concentration 10^{18} cm^{-3} and 100–200 Å thick with the adjacent wideband layer *n*-InGaAsP (E_g = 1.1 eV) and a thin *p*-layer 100–500 Å thick (Gorelenok et al. 1984). Such a base made it possible to reduce the transistor's dark current density from 10^{-4} to 10^{-6} A cm^{-2}. The increase of the thickness of the narrow-band

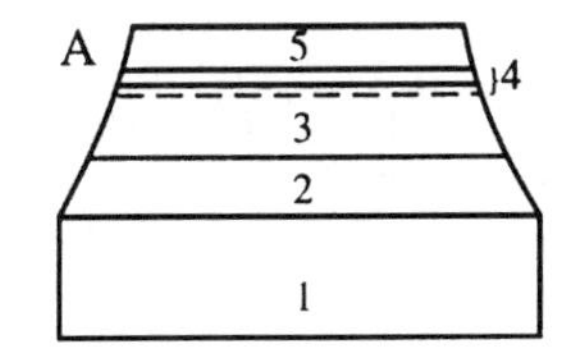

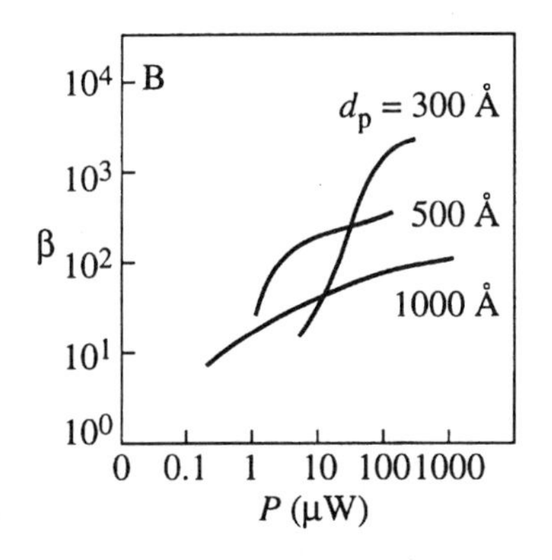

Figure 4.19. Bipolar *NpN*-phototransistor: (*A*) Schematic of the structure. (1) n^+-InP substrate; (2) InP buffer layer; (3) emitter (undoped InP, $n \sim 10^{17}$ cm^{-3}); (4) base (*p*-InGaAsP, E_g = 0.8 eV); (5) collector (InGaAsP, E_g = 1.1 eV, $n \sim 10^{17}$ cm^{-3}). (*B*) Dependence of current gain coefficient β on incident light power (P) for different thicknesses of the *p*-region d_p.

portion of the base from 100 to 1000 Å increases the current gain from 100 to 1000 (Fig. 4.19*B*) and the response time τ from 20 to 100 ns.

For the structures with different thicknesses of the base, a strong dependence of gain and speed on the incident light intensity was observed. This dependence is related to the barrier at the heterointerface. The maximum amplification $\beta \approx 1000$ with $\tau \sim 100$ ns was obtained at the incident light power $P \sim 10^{-4}$ W.

Two-Dimensional Electron Gas in InGaAs / InP Structures. The use of REE made it possible to fabricate InGaAs/InP two-dimensional electron gas structures by a liquid-phase epitaxy (Alferov et al. 1984). Later (Gorelenok et al. 1990) this technology was perfected. The $In_{0.88}Ga_{0.12}As_{0.23}P_{0.77}$ compound was used as a wideband gap material. In order to prevent the diffusion of iron from the substrate into the channel (Fig. 4.20, inset), a buffer InGaAsP layer was introduced between the substrate InP : Fe and

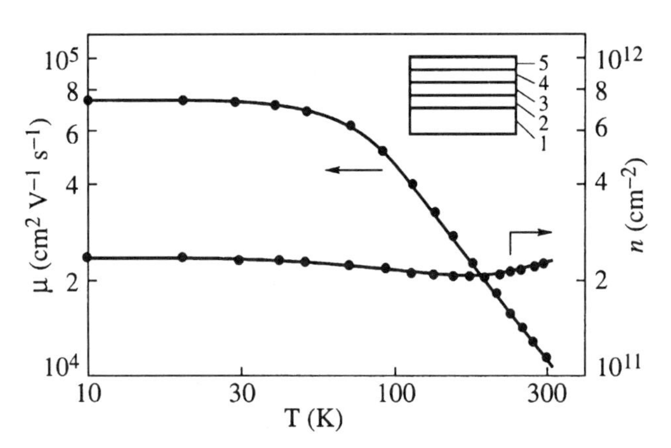

Figure 4.20. Temperature dependences of electron sheet concentration and mobility in two-dimensional electron gas. The $In_{0.88}Ga_{0.12}As_{0.23}P_{0.77}$/ $In_{0.53}Ga_{0.47}As$ structure with a buffer layer and a spacer is shown in the inset. (1) InP⟨Fe⟩ substrate; (2) n^0-InGaAsP ($\lambda_g = 1.06\ \mu$m) buffer layer, $n = 10^{14}$ cm^{-3}, $d = 0.5\ \mu$m; (3) InGaAsP ($\lambda_g = 1.06\ \mu$m) buffer layer, $n = 6 \times 10^{16}$ cm^{-3}, $d = 0.1$–$0.5\ \mu$m; (4) n^0-InGaAsP ($\lambda_g = 1.06\ \mu$m), $d = 100$–300 Å (spacer); (5) n^0-$In_{0.53}Ga_{0.47}As$, $n = 10^{15}$ cm^{-3}, $d = 0.4$–$3.0\ \mu$m.

the wideband gap layer. The electron concentrations in the narrow-band $In_{0.53}Ga_{0.47}As$ (0.7 μm thick) in the spacer (200 Å) and in the buffer layer (0.5 μm) of InGaAsP were reduced by REE doping to $n < 10^{15}$ cm^{-3}. The wideband gap layer of InGaAsP (0.2 μm thick) was undoped with and the electron concentration of 6×10^{16} cm^{-3}.

The results of the measurements in strong magnetic fields at 4.2 K are shown in Figure 4.21. These results and the analysis of the Shubnikov–de-Haas oscillations prove that the electron gas is two-dimensional.

The electron mobility in these structures was 12,300 cm^2 V^{-1} s^{-1} at 300 K and 7.3×10^4 cm^2 V^{-1} s^{-1} at 4.2 K. This is the best result for the structures obtained by LPE. It should be noted that in these structures the electron concentration and mobility do not depend on illumination (Gorelenok et al. 1991).

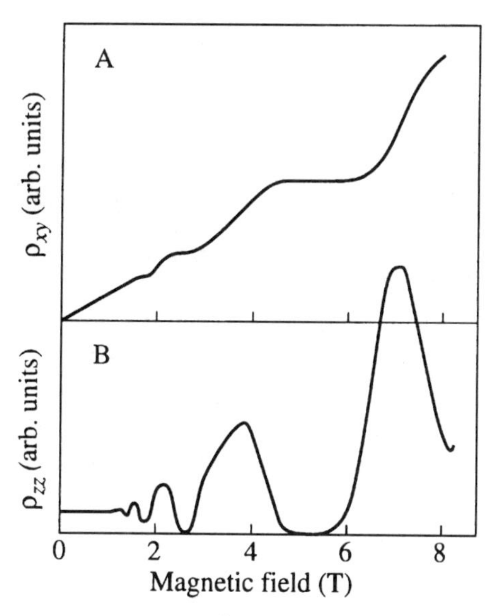

Figure 4.21. Quantum Hall effect (A) and Shubnikov-de Haas oscillations (B) in the two-dimensional electron gas for the structure shown in Fig. 4.20 at 4.2 K.

Schottky Barriers. The idea of applying REE for obtaining high-quality Schottky barriers was based on the assumption that the high chemical activity of REE to oxygen, arsenic, and phosphorus enables one to decrease the probability of the formation of unstable conducting phases of the InP and InGaAs native oxides. Also it was believed that REE would diminish the migration of As and P on the metal-InP(InGaAs) and insulator-InP(InGaAs) interfaces. This would result in a large Schottky barrier height, in a smaller density of the interface states, and in a better time stability of the device parameters.

To fabricate the InP and InGaAs Schottky barriers with the electron concentration $n = (1-2) \times 10^{16}\,\mathrm{cm}^{-3}$, a multilayer metallization was used: yttrium (200–300 Å thick), nickel (200–300 Å), and gold (5000 Å). A multilayer barrier was formed by a vacuum evaporation. The forward $I-V$ characteristics of the Schottky barriers were close to ideal with the ideality factor $n = 1.05-1.1$. The barrier height was 0.76 and 0.55 eV for InP and InGaAs, respectively.

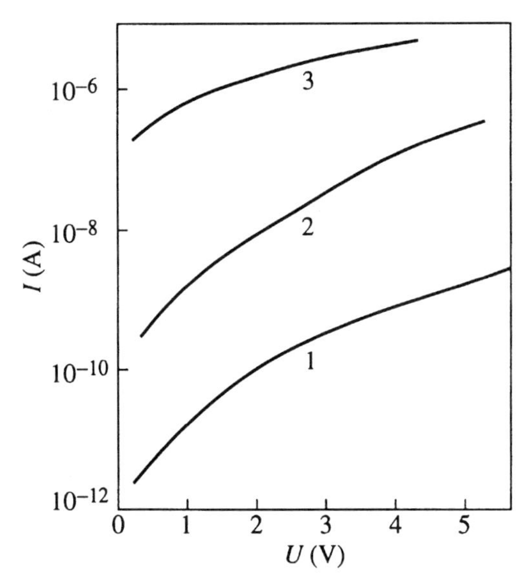

Figure 4.22. Reverse current-voltage characteristics of Schottky barriers at 300 K. The area is $9 \times 10^{-4}\,\mathrm{cm}^{-2}$. (1) InP–Yb–Ni–Au; (2) InGaAs–Yb–Ni–Au; (3) InP–Au.

Figure 4.22 shows the reverse I–V characteristics of these diodes (curves 1 and 2). For comparison, the I–V characteristic of Au–n-InP Schottky diode (curve 3) is also shown (Louliche et al. 1988). One can see that the leakage currents for the multilayer metallization at the reverse bias 1 V are lower by several orders of magnitude than those for the "classical" barriers Au–n-InP.

The preliminary investigations of the REE oxides showed that such rare-earth oxides as Sc_2O_3 and Y_2O_3 have a resistivity of 10^{15}–10^{16} $\Omega \cdot$ cm. These values are very close to the value of resistivity of thermal silicon dioxide. REE oxides were obtained by vacuum evaporation of scandium and yttrium with the controlled injection of oxygen. The investigation of InP-MIS (metal-insulator-semiconductor) structures demonstrated that the fixed interface charge was less than 10^{11} cm^{-2}, the surface state density in the minimum was less than 10^{11} eV^{-1} cm^{-2}, and the hysteresis of the C–V characteristics was smaller than 0.2 V (Belyakova et al. 1992).

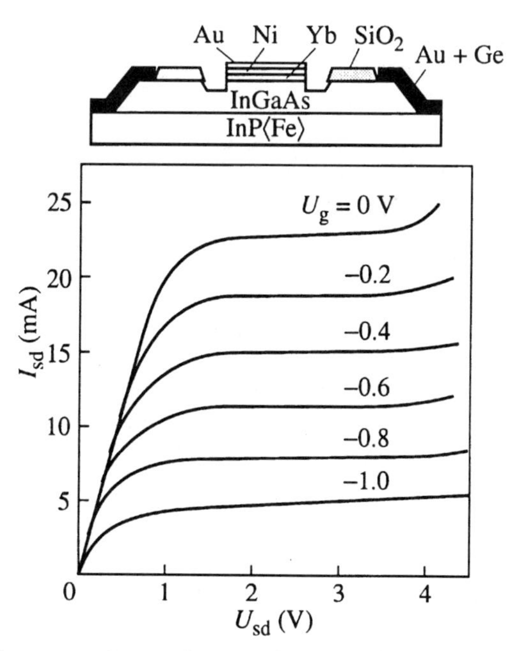

Figure 4.23. Current-voltage characteristics of InGaAsP/InP FETs with Schottky barrier gate at 300 K. The inset shows the schematic FET design.

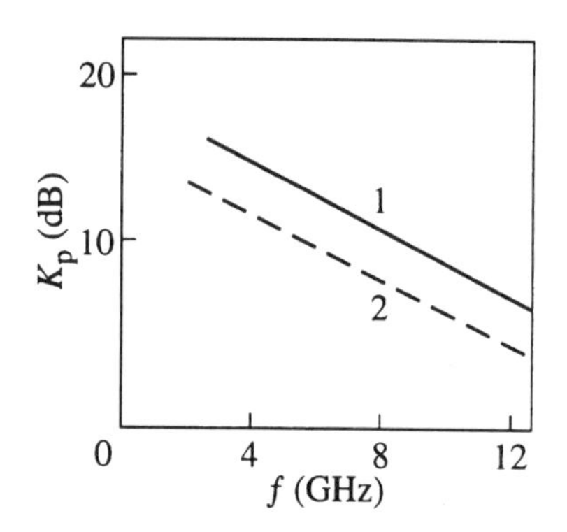

Figure 4.24. Frequency dependences of power gain at 300 K for Schottky gate FETs of InGaAs/InP (1) and of GaAs (2). The FETs were made under identical conditions and with the same design.

MESFETs. The technique of the simultaneous doping of Sn + REE was used to fabricate InGaAs MESFETs (Alferov et al. 1988). The InGaAs layers with the electron concentration $n = (1-2) \times 10^{17}\,\mathrm{cm}^{-3}$ had the electron mobility of $7000\,\mathrm{cm}^2\,\mathrm{V}^{-1}\,\mathrm{s}^{-1}$ at room temperature.

The schematic design and typical FET I–V characteristics are shown in Figure 4.23 (Alferov et al. 1988). For the gate dimensions of $1.5 \times 290\,\mu\mathrm{m}$, the transconductance was 180 mS/mm at 300 K and the peak power gain $K_p = 17$ dB was reached at 4 GHz. Figure 4.24 shows the frequency dependence of the power gain K_p.

For comparison, the same figure gives the frequency dependence (curve 2), measured for GaAs FET with the same design. One can see that InGaAs FET has a higher frequency of operation.

4.4. CONCLUSION

REE are very effective in the growth technology of GaAs and InP pure and semi-insulating single crystals, in the InGaAs and InP pure epilayers, and in related device technology. They allow us to grow materials and to fabricate devices with parameters as good as those obtained by more complicated and expensive methods that require highly pure initial materials.

The developed methods have potential for applications in industry. There is particularly much promise for the use of rare-earth oxides in coating various materials synthesized at high temperatures in aggressive media. However, more studies have to be done in order to understand the mechanism of the interaction of REE with semiconductor materials and their ability to control the defect formation in the semiconductor matrix.

REFERENCES

Afanas'ev, S. P., Karamov, F. G., and Bulekov, M. V. (1982). In *Technology of Semiconductor Compounds*, pp. 34–37. Metallurgy, Moscow.

Aldrich, C. and Greene, R. L. (1979). *Phys. Status Solidi* (*b*) 93, 343–350.

Alferov, Zh. I., Gorelenok, A. T., Danil'chenko, V. G., et al. (1983). *Sov. Techn. Phys. Lett.* 9, 649–650.

Alferov, Zh. I., Gorelenok, A. T., Kamanin, A. V., et al. (1984). *Sov. Phys. Semicond.* 18, 768–769.

Alferov, Zh. I., Bosyi, V. I., Gorelenok, A. T., et al. (1988). *Sov. Techn. Phys. Lett.* 14, 784–785.

Andreev, V. M., Bogdanovich, M. S., Gorelenok, A. T., et al. (1985a). *Sov. Phys. Techn. Phys.* 30, 907–909.

Andreev, V. M., Gorelenok, A. T., Zhingarev, M. Z., et al. (1985b). *Sov. Phys. Semicond.* 19, 411–414.

Bagraev, N. T., Bochkarev, E. P., Vlasenko, L. S., et al. (1978). *Inorg. Mater.* 14, 474–477.

Bagraev, N. T., Vlasenko, L. S., Gatsoev, K. A., et al. (1984). *Sov. Phys. Semicond.* 18, 49–50.

Bairamov, B. Kh., Zakharenkov, L. F., Il'menkov, G. V., et al. (1989). *Sov. Phys. Semicond.* 23, 927–929.

Balinas, V., Gorelenok, A. T., Krotkus, A., et al. (1990). *Sov. Phys. Semicond.* 24, 534–537.

Belyakova, E. D., Gabaraeva, A. D., Gorelenok, A. T. et al. (1992). *Poverkhnost'* 7, 88–93 (in Russian).

Bogdanovich, M. S., Korol'kov, V. I., Rakhimov, N., and Tabarov, T. S. (1985). *Sov. Techn. Phys. Lett.* 11, 35–36.

Bondina, M. I., Vavilov, E. V., Zakharenkov, L. F., and Sokolova, M. A. (1987a). Pat. USSR 1429612.

Bondina, M. I., Vavilov, E. V., Zakharenkov, L. F., and Masterov, V. F. (1987b). *Proc. I All-Union Conf. on Physical and Physical-Chemical Principles in Microelectronics*, Vilnius, USSR, 106 (in Russian).

Borisov, V. I., Gorelenok, A. T., Dmitriev, S. G., et al. (1992). *Sov. Phys. Semicond.* 26, 611–613.

Braudt, C. D., Hennel, A. M., Brysklewics, T., et al. (1989). *J. Appl. Phys.* 65, 3459–3469.

Deal, M. D. and Stevenson, D. A. (1984). *J. Electrochem. Soc.* 131, 2343–2347.

Ennen, H., Wagner, H. D., Muller, H. D., et al. (1987). *J. Appl. Phys.* 61, 4877–4882.

Evgen'ev, S. B. and Kuz'micheva, G. M. (1990). *Inorg. Mater.* 26, 1148–1151.

Factor, M. M. and Haigh, J. (1982). Pat. USA C30B29/40 N 4339302.

Fornari, R. (1991). *Mater. Sci. Eng.* B9, 9–18.

Galvanauskas, A., Gorelenok, A. T., Dobrovol'skis, Z., et al. (1988). *Sov. Phys. Semicond.* 22, 1055–1058.

Gatsoev, K. A., Gorelenok, A. T., Karpenko, S. I., et al. (1983). *Sov. Phys. Semicond.* 17, 1373–1375.

Golubev, V. G., Gorelenok, A. T., Ivanov-Omskii, V. I., et al. (1985). *Sov. Techn. Phys. Lett.* 11, 143–144.

Gorelenok, A. T., Gruzdov, V. G., Danil'chenko, V. G., et al. (1984). *Sov. Techn. Phys. Lett.* 10, 1294–1297.

Gorelenok, A. T., Danil'chenko, V. G., Dobrovol'skis, Z. P., et al. (1985). *Sov. Phys. Semicond.* 19, 1460–1463.

Gorelenok, A. T., Mamutin, V. V., Pulyaevskii, D. V., et al. (1987). *Sov. Phys. Semicond.* 21, 912–913.

Gorelenok, A. T., Gruzdov, V. G., Kumar Rakesh, et al. (1988). *Sov. Phys. Semicond.* 22, 21–26.

Gorelenok, A. T., Rekhviashvili, D. N., Nadtochii, M. Yu., and Ustinov, V. M. (1990). *Sov. Techn. Phys. Lett.* 16, 302–303.

Gorelenok, A. T., Rekhviashvili, D. N., Nadtochii, M. Yu., and Ustinov, V. M. (1991). *Sov. Phys. Semicond.* 25, 549–552.

Gorelenok, A. T., Kamanin, A. V., and Shmidt, N. M. (1995). *Microelectr. J.* 26, 705–723.

Gorelenok, A. T. and Shpakov, M. V. (1996). *Semicond.* 30, 269–271.

Jasiolek, G., Raczynska, J., and Gorecka, J. (1986). *J. Cryst. Growth* 78, 105–112.

Jasiolek, G. and Kalinski, Zb. (1989). *J. Cryst. Growth* 97, 583–586.

Jasiolek, G., Kalinski, Zb., Raczynska, J., and Paszkowics, W. (1989). *J. Mater. Sci.* 24, 2429–2432.

Karpov, Yu. A., Mazurenko, V. V., Petrov, V. V., et al. (1984). *Sov. Phys. Semicond.* 18, 230–231.

Körber, W., Weber, J., Hangleiter, A., et al. (1986). *J. Cryst. Growth* 79, 741–744.

Loualiche, S., L'Haridon, H., Le Corre, A., et al. (1988). *Appl. Phys. Lett.* 52, 540–542.

Mandelkorn, J., Schwartz, L., Broder, J., et al. (1964). *J. Appl. Phys.* 35, 2258–2260.

Markov, A. V., Bol'sheva, U. N., and Osvenski, V. B. (1993a). *High Purity Mater.* 1, 102–107 (in Russian).

Markov, A. V., Aref'ev, I. S., Bol'sheva, U. N., and Osvenski, V. B. (1993b). *High Purity Mater.* 1, 86–89 (in Russian).

Markov, A. V., Bol'sheva, U. N., Kardava, G. Ts., and Osvenski, V. B. (1993c). *High Purity Mater.* 1, 97–101 (in Russian).

Masterov, V. F., Savel'ev, V. P., Stel'makh, K. F., and Zakharenkov, L. F. (1989). *Sov. Phys. Semicond.* 23, 1381–1382.

Masterov, V. F. and Zakharenkov, L. F. (1990). *Sov. Phys. Semicond.* 24, 383–391.

Nakagome, H., Takahei, K., and Homma, Y. (1987). *J. Cryst. Growth* 85, 345–356.

Portnoi, K. I. and Timofeeva, N. I., eds. (1986). *Oxygen Compounds of Rare-Earth Elements*. Metallurgy, Moscow (in Russian).

Pyshkin, S. L., Radutsan, S. I., and Slobodchikov, S. V. (1967). *Sov. Phys. Semicond.* 1, 901–904.

Romanenko, V. N. and Kheifets, V. S. (1973). *Inorg. Mater.* 9, 172–177.

Selin, V. V. and Antonov, V. A. (1978). *Electronnaya Tekhnika, Ser. 6, Materials* 1, 57–62 (in Russian).

Stapor, A., Raczynska, J., Przybylinska, H., et al. (1986). *Mater. Sci. For.* 10–12, 633–638.

Stel'makh, K. F., Zakharenkov, L. F., Romanov, V. V., et al. (1990). *Sov. Phys. Semicond.* 24, 928–929.

Ulrici, W., Friedland, K., and Eaves, L. (1985). *Phys. Status Solidi (b)* 131, 719–728.

Volkov, L. A., Gorelenok, A. T., Luk'yanov, V. N., et al. (1987). *Sov. Techn. Phys. Lett.* 13, 442–443.

von Bardeleben, H. J. and Bourgoin, J. C. (1990). In *Defect Control in Semiconductors*, Sumino. K., ed., Elsevier, Amsterdam.

Zakharenkov, L. F., Kasatkin, V. A., Kesamanly, F. P., and Sokolova, M. A. (1981). *Sov. Phys. Semicond.* 15, 1631–1636.

Zakharenkov, L. F., Kasatkin, V. A., and Samorukov, B. E. (1983). Invent. Certificate of USSR 1085438.

Zakharenkov, L. F., Kasatkin, V. A., Litvin, A. A., and Makarenko, V. G. (1986). Invent. Certificate of USSR 1304441.

CHAPTER 5

INTRINSIC POINT DEFECT ENGINEERING IN SILICON HIGH-VOLTAGE POWER DEVICE TECHNOLOGY

N. A. SOBOLEV

The field of intrinsic point defect engineering started in the 1970s. It is based on the idea that nonequilibrium intrinsic point defects (IPDs) contribute to the formation of structural defects and electrically active centers as well as to the diffusion of impurity atoms. This has led to the development of different methods of gettering and passivation of defects in silicon devices.

Difficulties in studying the process of generation and relaxation of IPDs during the heat treatment are related to two factors. First, vacancies and self-interstitials cannot be observed as free single defects. They can be studied only by relying on the indirect data of structural defects, their electrically active centers, and the diffusion of doping impurities. Second, due to a low concentration of these defects, their nature cannot be investigated by direct methods such as electron paramagnetic resonance, optical absorption, etc.

Numerous publications (Hu 1974; Shiraki 1976; Ravi 1981; Antognetti et al. 1983; Taylor et al. 1989; Fahey et al. 1989) have dealt with the effects of IPDs on the formation of oxidation stacking faults and on the doping impurity diffusion. These studies

Semiconductor Technology: Processing and Novel Fabrication Techniques,
Edited by M. Levinshtein and M. Shur.
ISBN 0-471-12792-2

have provided a basis for the development of various methods of defect gettering and passivation.

Defect gettering and passivation techniques, developed for microelectronic devices, cannot be transferred to the technology of high-voltage power devices without conventional additional investigations, since the operating region for high-voltage power devices exceeds that for the microelectronic devices by several orders of magnitude. The operating region of a high-voltage device also is typically at sufficiently more depth from the device surface (~ 100 μm). The necessity of forming such deep p-n junctions requires rather long processes and higher temperatures. Besides, the silicon parameters used for manufacturing high-voltage devices differ a lot from those of silicon used in microelectronics. For manufacturing high-voltage devices, high pure dislocation-free *n*-float zone Si (*n*-FZ–Si) is used. This results in the concentrations of phosphorus and oxygen about two orders of magnitude lower than in the Czochralski silicon (Cz–Si) used in microelectronics. As far as we know, investigations of the generation and relaxation processes of IPDs in high-voltage devices have not been summarized earlier.

This chapter considers processes of generation and relaxation of IPDs in high-voltage power devices. In Section 5.1 we consider the effect of IPDs on the formation of the so-called swirl defects and on the deep diffusion of aluminium. Section 5.2 is devoted to the role of IPDs in forming the recombination centers in the neutron transmutation doped silicon (NTDS) and in forming deep-level centers (DLCs). Section 5.3 discusses the main principles of high-voltage power device technology.

5.1. EFFECT OF IPDs GENERATED UNDER HEAT TREATMENT ON THE FORMATION OF STRUCTURAL DEFECTS AND ON DOPING IMPURITY DIFFUSION

The type and concentration of IPDs can be controlled by changing the medium in which the heat treatment takes place (Fahey et al. 1989). The heat treatment in an oxidizing atmosphere results in the generation of self-interstitials; the heat treatment in an inert atmosphere and in a vacuum is accompanied by a supersaturation of silicon with vacancies, while the heat treatment in a chlorine-containing atmosphere makes it possible to control the type of dominating IPDs.

IPDs are formed under heat treatment in different media as a result of the chemical reactions at the surface of the silicon wafers. A well-known example of such reactions is the thermal oxidation of silicon:

$$Si + O_2 = SiO_2. \tag{5.1}$$

The oxidation leads to the supersaturation of the surface region with self-interstitials. This is conditioned by the constant lattice difference of Si and SiO_2. The mean distance between the silicon atoms in the oxide is about 1.3 times greater than in the silicon lattice. There are several models describing the process of generating excess self-interstitials during oxidation (Antognetti et al. 1983). However, there are not enough experimental data to check the adequacy of the proposed models.

During annealing in an inert atmosphere or in a vacuum, silicon is supersaturated with vacancies. This supersaturation is believed to be related to the chemical reaction at the $Si{-}SiO_2$ interface (the silicon surface is practically always coated with a thin layer of native oxide):

$$SiO_2 + Si = 2SiO. \tag{5.2}$$

The formed silicon monooxide diffuses through the thin oxide layer and evaporates into the vacuum or is carried away with the flow of the inert gas (Fahey et al. 1989). The decrease of silicon atom concentration is accompanied by the supersaturation of near-surface regions with vacancies. When the native oxide is fully depleted from the silicon surface via reaction (5.2), the generation of vacancies continues as a result of the evaporation of silicon atoms.

We must note that the above mechanism of vacancy generation has shown little efficiency during oxidation because the rate of decreasing the oxide thickness via reaction (5.2) is smaller by several orders than the growth rate of the silicon oxide via reaction (5.1).

The situation is more complicated when the silicon is annealed in a chlorine-containing atmosphere. This atmosphere presents a flow of oxygen saturated with the vapor of chlorine-containing compounds. The oxidation of the silicon surface in accordance with reaction (5.1) generates excess self-interstitials near the surface. The introduction of chlorine atoms into the atmosphere is accompanied by their accumulation in the SiO_2 layer near the $Si-SiO_2$ interface. On the one hand, that leads to an increase of the oxide growth rate and, consequently, to an increase of the silicon supersaturation with self-interstitials. On the other hand, the chemical reaction

$$xCl + Si = SiCl_x, \tag{5.3}$$

where $x = 1.0-2.1$ (Gresserov and Sobolev 1990) takes place on the $Si-SiO_2$ interface. The outdiffusion of chlorides, formed in the silicon surface regions, results in the generation of excess vacancies. Thus there are two competing processes that generate self-interstitials and vacancies. The annihilation of the defects-antipodes (vacancies and self-interstitials) leads to a decrease in the concentration of IPDs. In changing the annealing conditions (the concentration of the chlorine-containing component, time, temperature, etc.), it is possible to control the concentration of the excess IPDs.

5.1.1. Effect of Annealing Atmosphere on the Generation and Relaxation of IPDs during Annealing of Dislocation-Free FZ–Si Containing Swirl Defects

A high purity dislocation-free n-FZ–Si is used in the technology of high-voltage devices. The main type of structural defects in this type of silicon are microdefects formed from IPDs during single crystal growth. These defects may be microdefects of A, B, and other types (de Kock 1973), depending on the dimension, density, and the type of distribution under the selective chemical etching of silicon. High-resolution electron microscopy has showed that the larger A defects present dislocation loops of 1–5 μm in size of an interstitial type. These defects are formed as a result of the condensation of the excess interstitial atoms of silicon (Föll and Kolbesen 1975). The structure of the B defects, which are smaller, has not been established as yet, but there is every ground to believe that they are three-dimensional agglomerations of interstitial atoms of silicon that may include impurity atoms as well. A typical concentration of A and B defects is 10^6 and 10^7–10^8 cm^{-3}, respectively. The concentration may change depending on the growth conditions. The distribution of these defects in the cross section of crystals looks like a swirl, which explains their name "swirl defects."

The choice of swirl defects as an indicator, defining the type of IPDs generated during the annealing, was based on the results of d'Aragona (1971) and Ravi (1981). D'Aragona showed that the annealing of wafers with a polished surface in an argon atmosphere eliminates the swirl defects in the thin surface layer. Annealing in an oxidizing atmosphere is accompanied by a growth in their dimensions and concentration and even by their transformation into larger structural defects of an interstitial type—oxidation-stacking faults.

Sobolev et al. (1985) reported on the effect of an annealing atmosphere on the swirl defects within thick (1–5 mm) silicon wafers with a lapped surface. Dislocation-free *n*-FZ–Si crystals grown in the direction ⟨111⟩ were investigated. After selective etching, a distinct swirl distribution of microdefects was observed in the cross section of the crystals. According to the data on selective chemical etching, the concentration of microdefects was 4×10^6–6×10^8 cm^{-3} and the size of the etching pits was 3–10 μm.

According to the data obtained by the selective etching method, annealing in chlorine-containing atmosphere for 20–40 hours at 1100°–1250°C decreased the concentration of microdefects by the factor of 5 to 250. As the reduction took place, the dimensions of the etching pits on some samples were decreased by a factor of 2 or 3. The density of microdefects decreased with the increase of temperature and annealing time. In most of the investigated crystals, the macropattern of the swirl distribution of microdefects disappeared after the crystals had been annealed in the chlorine-containing atmosphere at 1250°C for 40 hours. The concentration of swirl defects after annealing in the air did not change or diminish even slightly, while the dimensions of the etching pits in practically all the samples increased by a factor of 2 to 4.

The decrease of the concentration of swirl defects during annealing in the chlorine-containing atmosphere is related to the generation of vacancies at the $Si-SiO_2$ interface via reaction (5.3), and their annihilation with self-interstitial atoms.

In the papers by Kurbakov et al. (1986) and Kurbakov and Sobolev (1994), γ-ray diffraction was used in order to investigate the behavior of swirl defects during annealing. The γ-irradiation with energy of 412 keV is characterized by weak absorption in silicon. This makes it possible to investigate large-diameter crystals. The rocking curves were measured in transmission geometry. This provided information on lattice disturbance along the thickness of the crystal in the direction of the incident beam. The integrated reflectivity R^{hkl} and the full width at half-maximum (FWHM) of the rocking curve served as the parameters characterizing the γ-ray diffraction for the reflection (hkl).

The γ-ray diffraction for starting samples revealed the presence of fields of tensile strain conditioned by swirl defects. The values of R^{111} were close to the maximum possible theoretical values (the so-called kinematic limit), with the half-width greatly exceeding the instrumental half-width. Annealing in the chlorine-containing atmosphere decreased both the integrated reflectivity and the half-width of the rocking curve compared to their values before annealing. This demonstrated the decrease of local stress in the bulk of the crystals, which is related to a decrease of the concentration of microdefects. Figure 5.1 shows the distributions of integrated reflectivity R^{111} measured along the axis of the crystal growth (A) and the rocking curves (B, C) for the specimen before annealing in the chlorine-containing atmosphere (A1, B) and after the annealing, which lasted 40 hours at 1250°C ($A2, C$). The values of the integrated reflectivity decreased by almost two orders of magnitude and became close to the minimum possible theoretical value (the so-called dynamical limit). The selective etching of the annealed crystal did not reveal a swirl pattern in the distribution of microdefects.

These results are of practical importance. Annealing in the chlorine-containing atmosphere makes it possible to diminish the concentration of as-grown microdefects with a swirl distribution within the thick silicon wafers. On the other hand, measuring the diffraction of high-energy γ rays opens new possibilities for the investigation of large monocrystals with nonuniform distributions of microdefects.

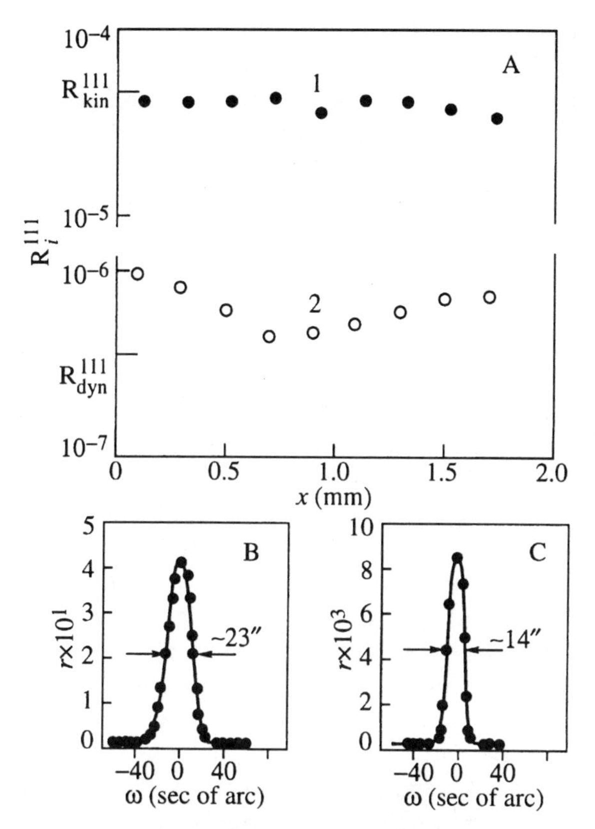

Figure 5.1. Variations of the integrated reflectivity (A) and rocking curves (B, C) for (111) reflection for n-FZ–Si. (A) Before annealing (1), after annealing (2); (B) before annealing; (C) after annealing. The annealing is carried out in a chlorine-containing atmosphere (T = 1250°C, t = 40 h, concentration of CCl_4 = 1.5 mol.%).

5.1.2. Effect of Annealing Atmosphere on the Generation and Relaxation of IPDs during Annealing of FZ–Si Irradiated by Neutrons

During the neutron transmutational doping of silicon, the crystals are irradiated by reactor neutrons with high doses. Simultaneously with the formation of doping atoms of phosphorus, radiation defects are created. Many publications have been devoted to the study of the effects of as-grown defects and of time and temperature of annealing on the structural and electrophysical properties of NTDS (e.g., Meese 1979; Ravi 1981). However, the effect of the annealing atmosphere on the formation of defects in FZ–Si after neutron irradiation has been studied much less. This section describes the results of an investigation on the effects of IPDs, generated by annealing radiation defects and/or chemical reactions (5.1)–(5.3), on the properties of FZ–Si irradiated by thermal neutrons.

A publication by Vysotskaya et al. (1988) has dealt with the investigation of structural defects in crystal that did not contain A defects before neutron irradiation and was annealed in different media at 850°C for four hours after irradiation. The defects were studied using X-ray topography by Lang (after the specimens were saturated with gold and copper) and by transmission electron microscopy (TEM). The presence of gold made it possible to reveal the A-type microdefects. X-ray topograms of the specimens showed that the parts of the crystal annealed both in a vacuum and in air contained defects nonuniformly distributed across the cross section. Near the edge of the crystal, the microdefect density was $10^5\ cm^{-3}$. In the central part the density decreased by two orders after annealing in air and practically to zero after annealing in a vacuum. In the specimens annealed in a chlorine-containing atmosphere, microdefects were not detected.

The structure of defects was studied by TEM. Single microdefects about 0.25 μm in size, presenting dislocation loops with the Burgers vector $(a/2)\langle 110 \rangle$ (A microdefects), were detected in the central part of the samples after annealing in a vacuum. Loops of about 0.2 μm in size, whose densities reached $10^6\ cm^{-3}$ as well as larger loops with a much smaller densities were detected at the peripheries of the samples. No microdefects were detected in the samples annealed in the chlorine-containing atmosphere. Complete dislocation loops of 0.45 and 1.1 μm in size and defects in the form of dislocation helicoids of about 2 μm in size comprising particles of α-crystabalite and α-quartz were detected after annealing in air. The density of helicoids was lower than $10^6\ cm^{-3}$.

The size and character of the microdefect distribution in the crystal bulk depend on the type of radiation defects, on the annealing medium, and on as-grown structural defects. The formation of the A-type microdefects in NTDS after annealing in a vacuum shows that the concentration of the excess self-interstitials generated during annealing is higher than the concentration of the excess vacancies generated via chemical reaction (5.2). In other words, when annealed in a vacuum, silicon becomes oversaturated with self-interstitial atoms. When the annealing is in air, the concentration of excess interstitials increases as a result of the chemical reaction (5.1). The nonuniform distribution of A defects in the bulk of the crystal after annealing of radiation defects is related to the presence of as-grown microdefects. After the initial crystal had been decorated with copper, the X-ray topograms showed the formation of microdefects in the periphery region. The character of the diffraction contrast did not allow one to determine the nature of these as-grown microdefects. The higher concentration of A defects near the edge of the crystal annealed in air and in a vacuum is conditioned by the as-grown microdefects, which present sinks for the excess interstitials.

The absence of A-type microdefects after annealing in a chlorine-containing atmosphere is caused by the generation of excess vacancies as a result of reaction (5.3) and then their annihilation with the excess interstitials according to the quasi-chemical reaction

$$V + I = \mathrm{Si_S}, \tag{5.4}$$

where V is a vacancy, I is a self-interstitial atom, $\mathrm{Si_S}$ is a substitutional silicon atom. These processes lead to the decrease of supersaturation of silicon with excess interstitials. Apparently prevention from external contamination and gettering of the impurity atoms from the bulk of specimens during chlorine annealing also facilitates obtaining NTDS defect-free crystals.

The investigation of structural defects in the FZ–Si crystals, irradiated by neutrons and annealed in different media, by the γ-ray diffraction technique, was described in Kurbakov et al. (1988) and Kurbakov and Sobolev (1994). A diffuse scattering caused by the defect formation was observed in the near-surface region of some crystals. After annealing in the argon atmosphere for four hours at 850°C, the diffuse scattering increased at the angles larger than the Bragg angle (Fig. 5.2*A*). This was associated with the formation of structural defects of an interstitial type. The formation of structural defects of an interstitial type observed in NTDS after annealing under similar conditions proved that at 850°C the mecha-

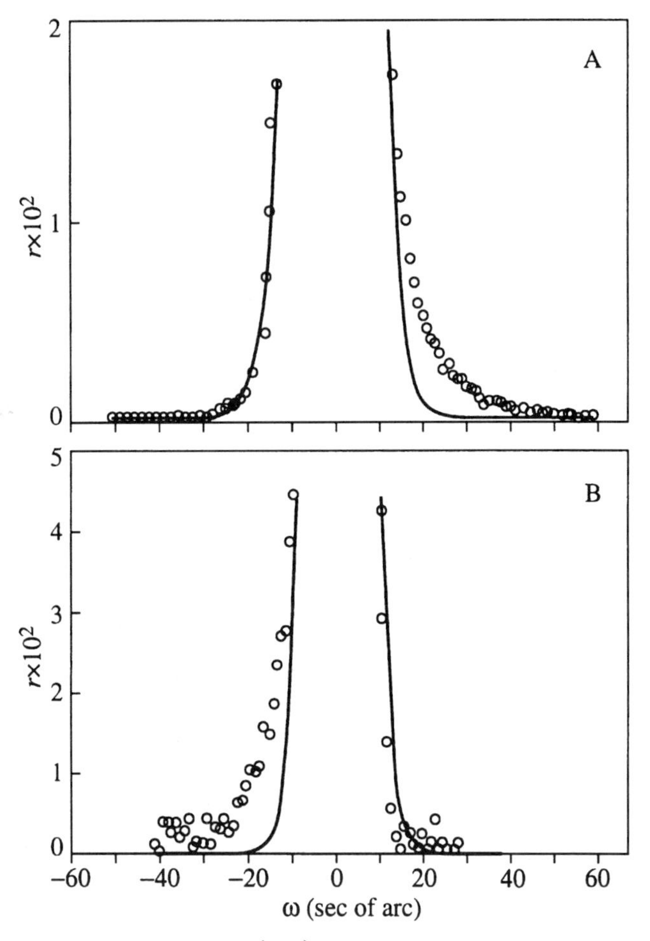

Figure 5.2. Rocking curves for (111) reflection of FZ samples irradiated by thermal neutrons and annealed (A) in an argon flow and (B) in a chlorine-containing atmosphere. Solid lines correspond to fitting by the Gauss function.

nism of generating vacancies via reaction (5.2) was not effective (Vysotskaya et al. 1988; Kurbakov and Sobolev 1994).

After annealing in the chlorine-containing atmosphere at 1050°C for four hours the diffuse scattering of γ rays was also discovered. As this took place, there was an increase of the diffuse scattering intensity at angles smaller than the Bragg angle (Fig. 5.2B), which is related to the formation of the structural defects of a vacancy type. The investigation of the same crystal by diffuse X-ray scattering (Sobolev et al. 1992) confirmed the defect formation of a vacancy type in the near-surface region. The angular dependence of diffuse X-ray scattering indicated the presence of defects of different sizes. The shape of the diffuse peak revealed that the defects were not spherical clusters. The diffuse γ-ray and X-ray scattering made it possible for the first time to observe "directly" the formation of the structural vacancy defects related to the generation of vacancies during silicon annealing in a chlorine-containing atmosphere.

Thus it was established that the use of a chlorine-containing atmosphere makes it possible to prevent the supersaturation of silicon with self-interstitials during postradiation annealing. There was discovered X-ray diffuse scattering from structural defects formed at participation of IPDs. For the first time the formation of vacancy defects during silicon annealing in a chlorine-containing atmosphere was observed by direct methods (by the diffusion scattering of γ and X-rays).

5.1.3. Effect of IPDs on the Diffusion of Aluminium in Silicon

The investigation of the effect of IPDs on the diffusion of impurity atoms allows one both to study the diffusion mechanism more thoroughly and to determine such important parameters of IPDs as the equilibrium diffusivity and the equilibrium concentration.

The diffusion of aluminium under the conditions of supersaturation of silicon with interstitials was observed to be accompanied by the increase of fabricated p-n junction's depth (Mizuo and Higuchi 1982). The excess self-interstitials were generated during the diffusion of aluminium into an unprotected silicon surface in the oxidizing atmosphere. The addition of chlorine-containing components into the oxidizing medium decreased the depth of the p-n junction compared to the diffusion in the oxidizing atmosphere (Mizuo and Higuchi 1982; Sobolev and Shek 1979). The experimental results were explained by assuming that the formed IPDs affected the diffusivity of aluminium in silicon (Mizuo and Higuchi 1982; Gösele and Tan 1983). In this case there was neglected the change of the concentration of the electrically active atoms of aluminium in the surface layer due to the generated IPDs.

In Mizuo and Higuchi (1982) aluminium was introduced by ion implantation into silicon wafers with polished surfaces. These conditions differed greatly from those used in the power device technology. Below we give the results of investigating the effect of IPDs on the diffusion of aluminium in silicon under conditions typical for high-voltage power device fabrication (Gresserov et al. 1991).

The aluminium diffusion was investigated at 1250°C. For the process to be conducted when the silicon was supersaturated with excess vacancies or self-interstitials, the diffusion was performed in nitrogen and in oxygen, respectively. The NTDS wafers with a resistivity of 140 $\Omega \cdot$ cm were lapped with the SiC micropowder with grain size of 20 μm. Films containing aluminium nitrate were used as an aluminium source. The concentration profiles were determined by the differential conductivity technique. The concentration dependence of the mobility was taken from the publication (Caughey and Tomas 1967).

The investigation of aluminium concentration profiles under diffusion for 10–40 hours in different media has shown the following:

- The concentration profiles after the diffusion in oxygen are described by the Gauss function.
- The concentration profiles after the diffusion in nitrogen have a nonclassical form and cannot be adequately described by the Gaussian curve.
- Under diffusion in nitrogen the surface concentration of aluminium is 1.5 to 2 orders of magnitude higher, and the depth of the p-n junction is 20%–30% smaller, than for diffusion taking place in an oxidizing medium.

Figure 5.3 shows the typical aluminium profiles after diffusion in inert and oxidizing media. Experimental results can be qualitatively explained in the following way: Diffusion in nitrogen is accompanied by the generation of excess vacancies. Their interaction with

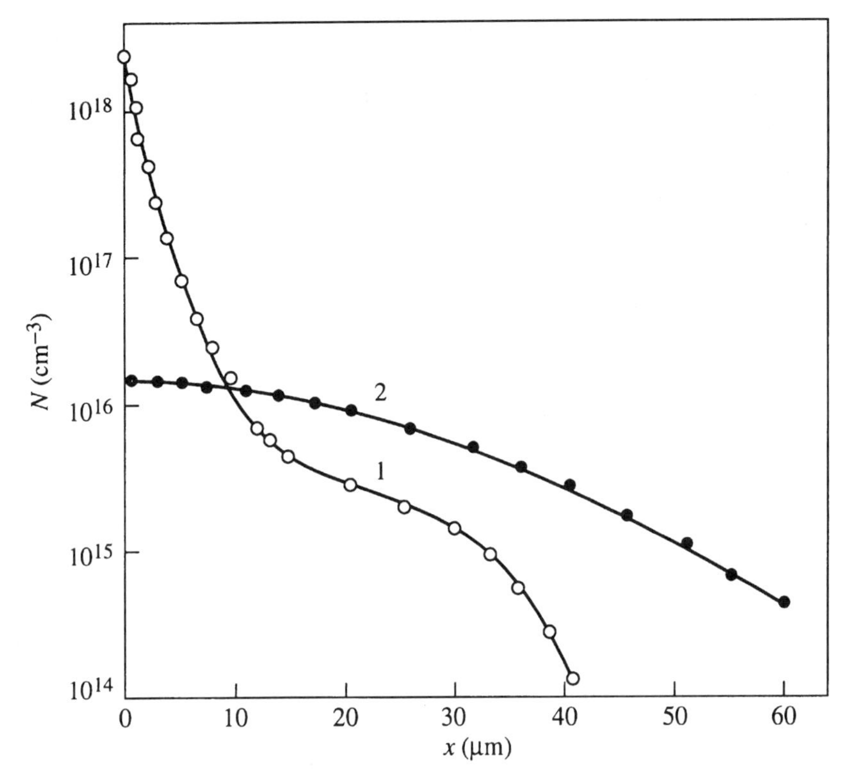

Figure 5.3. Concentration profiles of Al in Si after diffusion (T = 1250°C, t = 16 h) in nitrogen (1) and in oxygen (2). The points are the experimental data, and the curves are calculated.

the aluminium atoms can be described by the reaction

$$Al_I + V = Al_S, \tag{5.5}$$

where Al_I and Al_S are the aluminium atoms in their interstitial and substitutional positions. That leads to an increase in the concentration of the electrically active aluminium atoms in the near-surface region. However, since the diffusion of aluminium in silicon takes place mostly according to the interstitialcy mechanism (Mizuo and

Higuchi 1982), the undersaturation of silicon by self-interstitials during the diffusion in nitrogen is accompanied by the decrease of the effective diffusivity and, accordingly, by the decrease of the p-n junction depth compared to the diffusion in oxygen.

The analysis of experimental results showed that the diffusion of aluminium in silicon can be described by the model based on the following assumptions:

- Interaction of self-interstitials and vacancies in the bulk of the sample can be maintained by quasi-chemical reactions (5.4).
- Different charge states of IPDs can be ignored.
- The internal sources and sinks for the interstitials and vacancies are lacking, and only the sample surface can serve as their source.
- Interaction of the aluminium atoms and IPDs can be determined by introducing an effective diffusivity that depends on the IPDs' concentration:

$$\frac{\partial V}{\partial t} = D_V \frac{\partial^2 V}{\partial x^2} - k_0(IV - I^* V^*), \tag{5.6}$$

$$\frac{\partial I}{\partial t} = D_I \frac{\partial^2 I}{\partial x^2} - k_0(IV - I^* V^*), \tag{5.7}$$

$$\frac{\partial N}{\partial t} = \frac{\partial}{\partial x}\left(D_N \frac{\partial N}{\partial x} \right), \tag{5.8}$$

$$D_N = D^* \left[f\left(\frac{I}{I^*}\right) + (1 - f)\left(\frac{V}{V^*}\right) \right]. \tag{5.9}$$

Here I and V are the concentration of interstitials and vacancies, respectively (the thermodynamically equilibrium values are denoted by asterisks), D is diffusivity, N is the dopant concentration, k_0 is a constant, and f is a coefficient. It should be noted that according to capacitance spectroscopy, the local dynamical equilibrium between the vacancies and interstitials has not yet been reached under our experimental conditions [i.e., the mass action law $IV = I^*V^*$ does not hold; see (5.6) and (5.7)]. The second factor in (5.9) reflects the contribution of IPDs to the diffusion of interstitial aluminium atoms. The first term in the square brackets represents the contribution of the indirect interstitial diffusion mechanism. In this case the self-interstitials push out the impurity atoms into the interstitial positions according to the following reactions (Fahey et al. 1989):

$$\mathrm{Al}_S + I = \mathrm{Al}I, \quad \mathrm{Al}_S + I = \mathrm{Al}_I. \tag{5.10}$$

Then the impurity atom occupies a position in the nearest site in accordance with reaction (5.5). The second term in the square brackets represents the contribution of the vacancy exchange mechanism, namely the movement of the substitutional atoms jumping to the nearest substitutional positions. The coefficient f describes the contribution of the interstitialcy mechanism to the effective diffusivity of aluminium: $0.5 < f < 1$ when the interstitialcy mechanism is dominant, and $0 < f < 0.5$ when the vacancy mechanism is dominant.

Numerical analysis of the experimental results has shown that aluminium diffusion in silicon at 1250°C is controlled by the concentration of self-interstitials ($f > 0.95$) and that the thermodynamically equilibrium diffusivity is equal to 2×10^{-11} cm^2/s. That value of self-interstitial diffusivity corresponds to values determined experimentally from boron diffusion in silicon (Gossmann et al. 1995). The observed decrease of aluminium penetration depth during diffusion in nitrogen, compared to that in oxygen, was caused by the decrease of effective diffusivity as a result of the vacancy supersaturation in the near-surface layer of silicon.

Thus experimental study of the behavior of structural defects and of diffusion of aluminium in silicon has revealed an essential role of IPDs in defect formation and has demonstrated the possibility of regulating IPD generation under conditions typical for high-voltage power devices. The next section is devoted to the effect of IPDs on the formation of electrically active centers.

5.2. EFFECT OF IPDs GENERATED UNDER HEAT TREATMENT ON THE FORMATION OF ELECTRICALLY ACTIVE CENTERS

5.2.1. Effect of IPDs on the Formation of Electrically Active Centers in Neutron Transmutation-Doped Silicon

The effect of IPDs formed under postradiation annealing on the formation of electrically active centers has been investigated by Voronov et al. (1984), Kurbakov et al. (1988), and Sobolev et al. (1991). IPDs were formed as a result of annealing radiation defects and chemical reactions (5.1)–(5.3) proceeding on a crystal surface during annealing in different media. For the crystals irradiated by neutrons with a fluence of thermal neutrons of $\sim 10^{17}\,\text{cm}^{-2}$ and with a ratio of thermal and fast neutrons of 20 : 1, the annealing medium does not affect the resistivity of NTDS, at least within $\pm 5\%$ for the values of resistivity of 40–350 $\Omega \cdot \text{cm}$. At the same time the lifetime of minority charge carriers depends on the annealing atmosphere and increases in the following order: argon, vacuum, air, and chlorine-containing atmosphere.

The annealing of radiation defects in a chlorine-containing atmosphere makes it possible to obtain NTDS with minority carrier lifetimes that are practically the same as for the crystals before their irradiation with neutrons. Annealing in vacuum and in air leads to a decrease of lifetime two to three times that of chlorine annealing; the values of lifetime are practically independent of their values in the starting crystals and are determined by the annealing conditions.

The formation of the A-type microdefects in NTDS after the annealing in argon, a vacuum, and air (see Section 5.1.2.) has led us to conclude that self-interstitials play an important role in forming the recombination centers. Roughly equal lifetimes after the annealing in vacuum and in the air indicate that the supersaturation of silicon with self-interstitials is determined to a great extent by the conditions of neutron irradiation. A relatively small difference in the lifetimes after annealing in these media can be explained by the participation of impurity atoms of different type in the formation of recombination centers.

The lack of interstitial structural defects and the preservation of lifetime at the same level as for the starting crystals testify that the annealing of radiation defects in a chlorine-containing atmosphere prevents the formation of recombination centers of the interstitial type.

The distribution of lifetimes along the depth of the crystal was investigated by Kurbakov et al. (1988). After annealing in a chlorine-containing atmosphere for four hours at 1050°C, an essential increase of lifetime in the near-surface region was observed. This increase can be explained as follows. Just like for crystals annealed in argon, a vacuum, and air, one can assume that in the bulk of a crystal, annealed at 1050°C, the process of recombination of nonequilibrium carriers is controlled by the interstitial-type defects. The generation of vacancies under annealing in a chlorine-containing atmosphere results in a decrease of the interstitial recombination center concentration and in an increase of lifetime in the near-surface region. It should be noted that the formation of the vacancy structural defects was observed in the near-surface region of the same crystal using γ-ray and X-ray diffuse scattering (see Section 5.1.2). These data demonstrate an important role played by IPDs in forming recombination-generation centers in NTDS.

5.2.2. Effect of IPDs on the Formation of Deep-Level Centers

This section summarizes the results of work on deep-level center (DLC) formation which is related to the generation and relaxation of IPDs under the heat treatment of silicon crystals and structures with p-n junctions at high temperatures $> 1000°C$ (Sobolev et al. 1991; Vyzhigin et al. 1991). DLCs were investigated in the dislocation-free n-FZ–Si and n-Cz–Si with the resistivity 10–350 $\Omega \cdot$ cm. To generate IPDs, wafers with polished surfaces were annealed, or aluminium was diffused into wafers with lapped surfaces in different media at 1000°–1250°C for 1 to 16 hours. To measure the DLC parameters in the Schottky barriers or p-n junctions, different methods of capacitance spectroscopy were used such as deep-level transient spectroscopy (DLTS), isothermal capacitance transient, thermally stimulated capacitance, and photocapacitance.

Comparing the data for the samples after heat treatment in oxygen, one can see that at every temperature-time regime that was investigated, two dominating deep levels $E1$ and $E4$ are formed. For the $E1$ center the thermal emission rate of electrons to the conductance band is as follows:

$$e_1 = 1.6 \times 10^{-15}\, bT^2 \exp\left(-\frac{0.535}{kT}\right). \tag{5.11}$$

For the $E4$ center

$$e_4 = 1.9 \times 10^{-17}\, bT^2 \exp\left(-\frac{0.277}{kT}\right). \tag{5.12}$$

Here $b = 6.6 \times 10^{21}\,\text{cm}^{-2}\,\text{c}^{-1}\,\text{K}^{-2}$, T is the temperature in K, k is the Boltzmann constant, e was measured within the range 10^{-2}–$10^{3}\,\text{s}^{-1}$. Depending on the technological conditions, the concentrations of the centers were $M4 = 10^{11}$–$10^{12}\,\text{cm}^{-3}$ and $M1 = (0.6–0.9)M4$.

The heat treatment of samples in an inert atmosphere or in a vacuum is accompanied by the formation of three DLCs with the following rates of thermal emission:

$$e_3 = 1.2 \times 10^{-14}\, bT^2 \exp\left(-\frac{0.455}{kT}\right), \tag{5.13}$$

$$e_5 = 4.0 \times 10^{-16}\, bT^2 \exp\left(-\frac{0.266}{kT}\right), \tag{5.14}$$

$$e_7 = 1.1 \times 10^{-15}\, bT^2 \exp\left(-\frac{0.192}{kT}\right). \tag{5.15}$$

The concentration of the dominant center ($E5$ or $E7$) is typically within 10^{12}–$10^{13}\,\text{cm}^{-3}$, depending on the technological conditions. The $E3$ center concentration is 10 to 50 times lower.

The heat treatment at a rather high temperature in the oxidizing atmosphere is accompanied by supersaturation of Si with self-interstitials, while the heat treatment in an inert medium or in a vacuum leads to the vacancy supersaturation (see Section 5.1). This shows that the interstitials may play a major role in the formation of DLCs' $E1$ and $E4$ levels, while the vacancies play a decisive role in the formation of $E3$, $E5$, and $E7$ levels (Sobolev et al. 1989).

Some additional experiments were carried out to confirm the above hypothesis of the IPDs' role in forming the $E1$, $E3$–$E5$, and $E7$ centers.

Simultaneous measurements by γ-ray diffraction and capacitance spectroscopy were made on a number of samples annealed in different media (Sobolev et al. 1991). The $E1$ and $E4$ centers dominated in the DLTS spectrum in those samples in which, according to the data of diffraction experiments, the formation of structural defects of an interstitial type was observed. The $E3$, $E5$, and $E7$ centers dominated in those samples where the structural defects of a vacancy type were revealed. Depending on the conditions of heat treatment, the relation between the concentrations of various DLCs in the samples under investigation changed by more than an order of magnitude. A comparison of the data of diffraction and capacitance measurements directly confirmed that the formation of the $E1$ and $E4$ centers occurred under conditions of supersaturation of Si with excess interstitials, while the formation of the $E3$, $E5$, and $E7$ centers took place under conditions of supersaturation of Si with vacancies. But we cannot state if the DLCs under investigation belonged to the same structural defects that were revealed by the γ-ray diffraction.

Figure 5.4 shows the DLTS spectra after aluminium diffusion in nitrogen (A) and oxygen (B) atmospheres. The aluminium impurity concentration profiles for the same samples are shown in Figure 5.3. The analysis of the concentration profiles above has shown that aluminium diffusion in a nitrogen atmosphere proceeds under the supersaturation of silicon with vacancies and with self-interstitials in oxygen. This proves that vacancies play a crucial role in the formation of $E3$, $E5$, and $E7$ centers, while interstitial silicon atoms play a similar role in the formation of $E1$ and $E4$ centers.

Under the diffusion of aluminium in silicon in an inert atmosphere, an $E4$ level is formed in addition to the $E3$, $E5$, and $E7$ levels. The simultaneous formation of both vacancy and interstitial DLCs is related to the coexistence of IPDs at high temperatures; the generation of vacancies is determined by chemical reaction (5.2)

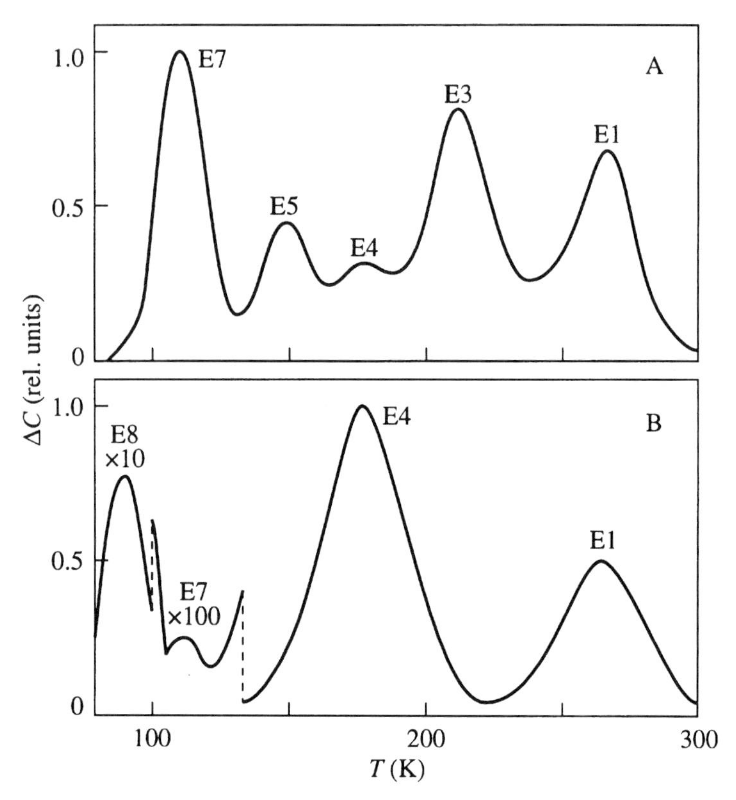

Figure 5.4. DLTS spectra of samples after the diffusion of Al in Si (A) in nitrogen and (B) in oxygen (T = 1250°C, t = 16 h). The emission rate of carriers from a level at the peak maximum was 60 s^{-1}.

proceeding under heat treatment in an inert atmosphere, while the generation of self-interstitial atoms takes place by the reaction

$$Al_i + Si_s = Al_s + I, \tag{5.16}$$

where i and s show that the atom is in an interstitial position or in a lattice site, respectively.

Under aluminium diffusion in silicon in a chlorine-containing atmosphere, the concentration of $E1$ and $E4$ levels decreases compared to diffusion in oxygen, and the formation of an $E7$ level is observed (Sobolev et al. 1989). The change of the DLTS spectrum is due to the generation of vacancies at the $Si{-}SiO_2$ interface via reaction (5.3) and with the corresponding decrease of the supersaturation of silicon with self-interstitials.

The DLTS spectra of the p-n structures fabricated by the aluminium diffusion in the argon atmosphere (A) and subjected to the subsequent annealing in oxygen (B) are shown in Figure 5.5. In Figure 5.5 the *C-E* curves show the spectra of the p-n structures fabricated by aluminium diffusion in an oxygen atmosphere (C) and then annealed in argon (D) or in a vacuum (E). The diffusion and annealing took place at 1250°C (in a vacuum at 1230°C) for four hours. The change of the DLTS spectrum in accordance with the type of IPDs generated during annealing is an important argument in favor of the proposed model of the formation of $E1$, $E3{-}E5$, and $E7$ centers. The different concentrations of DLCs that are formed under repeat annealing are conditioned by different rates of the generation of vacancies for annealing in argon or in a vacuum (curves D and E in Fig. 5.5).

Heat treatment of FZ and Cz–Si under identical conditions results in the formation of similar spectra of DLCs. This means that the oxygen impurity does not affect the formation of centers, since the starting concentration of oxygen in the FZ and Cz–Si differs by more than an order of magnitude.

It is important to note that different defects can have the same temperature dependence of the thermal emission rate. Hence it may be necessary to conduct additional experiments to identify the defects. In our case the temperature dependence of the electron emission rate for the $E1$ level coincided with these dependencies for the levels belonging to the impurity atoms of Au, S, Ag, Co, and Mn in n-Si (Chen and Milnes 1980; Berman and Lebedev 1981). To obtain additional information, the effect of the hydrostatic pressure on the parameters of the $E1$ center was investigated (Vyzhigin et al. 1989). Unlike the data for the level of gold shown by Samara and Barnes (1987), the hydrostatic pressure coefficient determined by us for the $E1$ level was almost twice as small and practically did not depend on temperature (it was equal to $13.9\,\mathrm{meV/GPa}$ within the temperature range 234–291 K); that is to say, the $E1$ level is not the level of gold. The $E1$ level does not belong to the defect related to the sulfur atoms either, since the values of the hydrostatic pressure

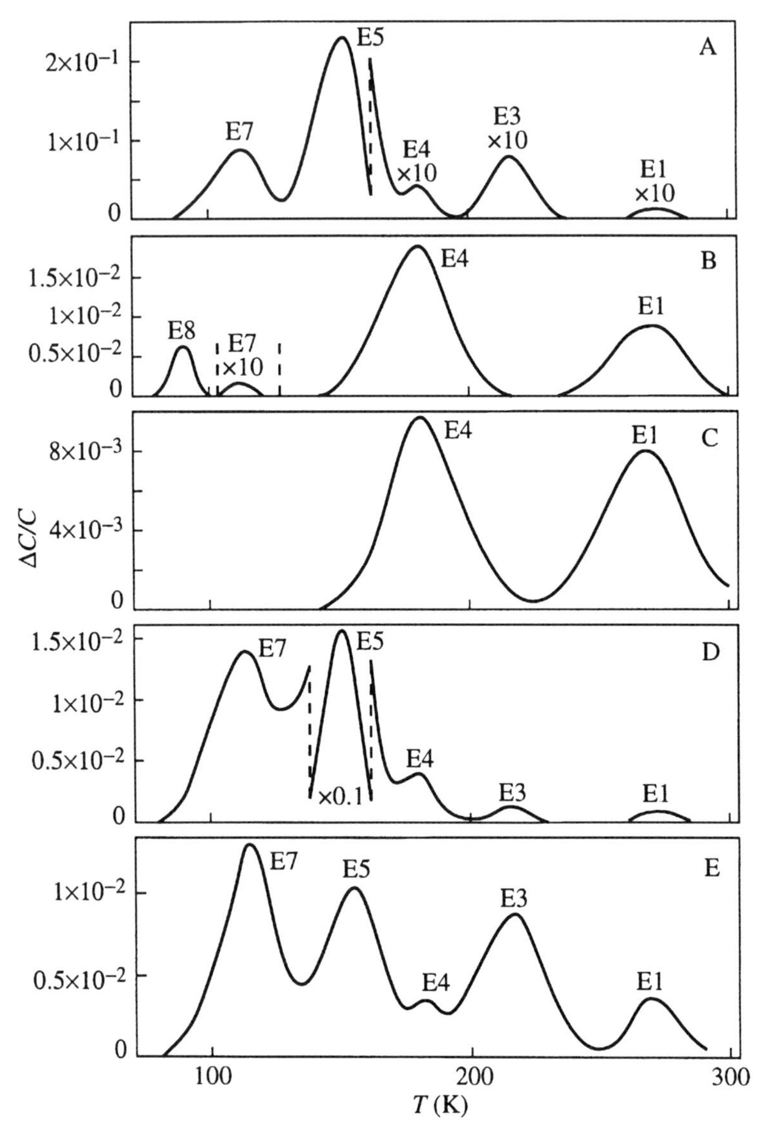

Figure 5.5. DLTS spectra of samples after Al diffusion (A) in argon and (C) in oxygen and after subsequent annealing (B) in oxygen, (D) in argon, and (E) in vacuum.

coefficient for the sulfur and $E1$ level are also quite different (Jantsch et al. 1982).

The formation of DLCs, characterized by the dependence $e(T)$ and practically coinciding with the centers under investigation, were observed in quite a number of publications (see Table 5.1). The formation of identical centers under different conditions (types of

TABLE 5.1. Parameters of DLCs formed in Si after heat treatment in different media.

Range of Thermal Emission Rate (s^{-1})	Notation of the Center Activation energy (eV) Electron Capture Cross Section (cm^2)					References
			Heat treatment in oxidizing atmosphere			
10^{-2}–10^{3}	M 0.542 2.5×10^{-15}		U 0.264 9.3×10^{-18}			Yah and Sah (1974)
10^{-2}–10^{0}	M 0.538 2.4×10^{-15}		U 0.276 2.6×10^{-17}	X 0.298 5.2×10^{-15}	Y 0.222 2.8×10^{-15}	Sah and Wang (1975)
10^{-2}–10^{3}	$E1$ 0.54 2.5×10^{-15}		$E2$ 0.27 1.5×10^{-17}			Berman and Lebedev (1981)
$(2$–$10) \times 10^{2}$	M 0.54 1.0×10^{-15}		U 0.28 4.0×10^{-17}	L 0.34 6.0×10^{-14}		Astrova et al. (1987)
10^{-2}–10^{3}	$E1$ 0.535 1.6×10^{-15}		$E4$ 0.277 1.9×10^{-17}		$E7$ 0.192 1.1×10^{-15}	Vyzhigin et al. (1991)
			Heat treatment in vacuum or inert atmosphere			
10^{0}–2×10^{2}	260 0.561 4.6×10^{-15}	204 0.460 2.3×10^{-14}	175 0.264 9.3×10^{-18}	145 0.356 5.9×10^{-13}	106 0.236 5.6×10^{-14}	Paxman and Whight (1980)
10^{0}–8×10^{2}				0.25 1.3×10^{-16}		Senes (1981)
5.6×10^{2}					X_2 0.19 1.15×10^{-15}	Kimerling et al. (1981)
$(1.5$–$3) \times 10^{2}$	$A1$ 0.54 1.5×10^{-15}		C 0.31 8.6×10^{-17}	D 0.31 2.7×10^{-15}	E 0.26 3.9×10^{-14}	Crees and Taylor (1982)
10^{-2}–10^{3}		$E3$ 0.455 1.2×10^{-14}	$E4$ 0.277 1.9×10^{-17}	$E5$ 0.266 4.0×10^{-16}	$E7$ 0.192 1.1×10^{-15}	Vyzhigin et al. (1991)

silicon and technological regimes) in the research laboratories in different countries is an additional argument in favor of the fact that the formation of these defects is caused by IPDs.

5.2.3. Effect of IPDs on the Formation of Electrically Active Centers in Structures with p-n Junctions

This section contains the results of the investigation of the effect of IPDs on the formation of electrically active centers, controlling the processes of generation, recombination, and impact ionization in deep p-n junctions (Kostylev et al. 1989; Sobolev et al. 1991; Vyzhigin et al. 1992).

P-n junctions based on n-Si with resistivity 45–300 $\Omega \cdot$ cm were investigated. The p-n junctions were obtained by diffusion of aluminium (with surface concentration $\sim 10^{17}\,\mathrm{cm}^{-3}$) into lapped wafers. The depths of the p-n junctions were 120–140 μm. High doped p^+ and n^+ layers were obtained by boron (10^{17}–$10^{20}\,\mathrm{cm}^{-3}$) and phosphorus ($\sim 10^{20}\,\mathrm{cm}^{-3}$) doping. Diffusion and annealing proceeded at 1000°–1250°C for 40 minutes to 40 hours in different atmospheres. In the structures under investigation, DLCs of an interstitial or vacancy type were dominant. Their concentrations varied in the range of 10^{10}–$10^{12}\,\mathrm{cm}^{-3}$. The minority carrier lifetimes were 10–400 μs.

The minority carrier lifetime was determined by the duration of the high reverse conductance after the p-n structure was switched from a conducting to a nonconducting state (Lax and Neustadter 1954). To measure the bulk component of the reverse current, a guard ring technique was used (Volle et al. 1975). The parameters of the centers responsible for the microplasmas were determined by microplasma spectroscopy technique (Vyzhigin et al. 1988).

The DLC parameters and lifetime were measured in structures made by the diffusion of aluminium. The diffusion was performed at 1250°C first for 4 hours in a flow of argon and then for 4 to 32 hours in oxygen. The atmosphere was changed without switching off the furnace at the diffusion temperature (Vyzhigin et al. 1992).

Figure 5.6 gives the dependence of the total center concentration M of both vacancy and interstitial types on the duration of annealing in oxygen. These concentrations are normalized to the shallow donor concentration N. In the first stage (diffusion in argon), excess

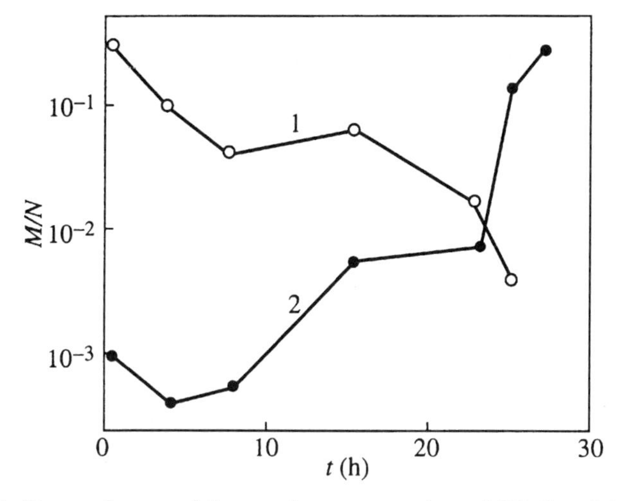

Figure 5.6. Dependence of the total concentration of DLCs of the vacancy (1) and interstitial (2) types on the duration of annealing in an oxygen flow.

vacancies are mostly generated. If at that moment the sample is cooled to the room temperature, those excess vacancies lead to the formation of vacancy centers $E3$, $E5$, and $E7$. In the second stage (diffusion in oxygen), self-interstitials of Si are mostly generated, and their bimolecular recombination with the vacancies (which were formed in argon under diffusion) takes place. With an increase of duration in the second stage, the concentration of the excess vacancies decreases, while the concentration of self-interstitials increases. After some time the type of dominating IPDs and DLCs changes.

The dependence of lifetime on the duration of diffusion in oxygen is shown in Figure 5.7. The maximum in this dependence can be explained in a similar way. The lifetime after diffusion in argon is determined by the vacancy recombination centers. The diffusion in oxygen is accompanied by the generation of the self-interstitials. The bimolecular recombination of the interstitials with the excess vacancies (which appear under the diffusion in oxygen) leads to a decrease in the concentration of the recombination centers and to a lifetime increase. With an increase in the duration of diffusion in oxygen, the type of dominating IPDs changes, and

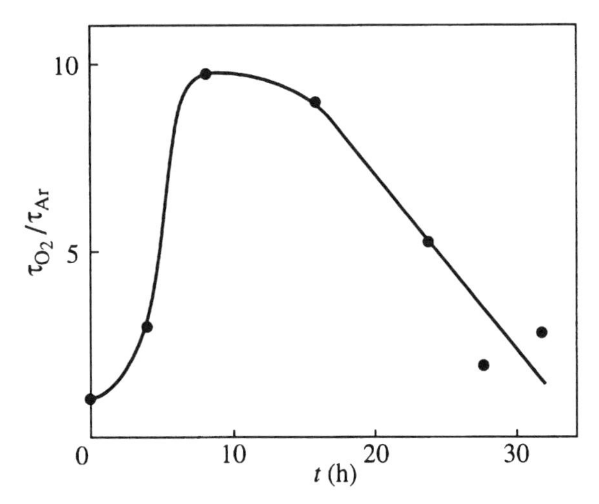

Figure 5.7. Dependence of the minority carrier lifetime on the duration of annealing in an oxygen flow. Here τ_{Ar} and τ_{O_2} are the lifetimes measured after the diffusion of Al in the argon atmosphere and after the subsequent annealing in oxygen, respectively.

recombination centers of an interstitial type are formed. As their concentration increases, they begin to control the recombination process of nonequilibrium charge carriers; this leads to a drop in the lifetime.

Comparing Figures 5.6 and 5.7, one can see that the positions of the minimum of the total concentration of DLCs and the maximum of the lifetime do not coincide. This may be explained by the fact that the lifetime is determined by recombination via several centers with different capture cross sections of carriers.

It is impossible to describe the recombination of nonequilibrium charge carriers in the investigated structures within the framework of a one-level generation-recombination Shockley-Read model, even though the tendency of lifetime increase with the decrease of concentration of DLCs is observed.

The parameters of the centers controlling the volume component of the reverse current were investigated in p-n structures containing DLCs of the interstitial type $E1$ and $E4$ (Kostylev et al. 1989). It turned out that the volume component of the reverse current in the structures with a concentration of centers $E1$ and $E4$ $(0.5-2) \times$

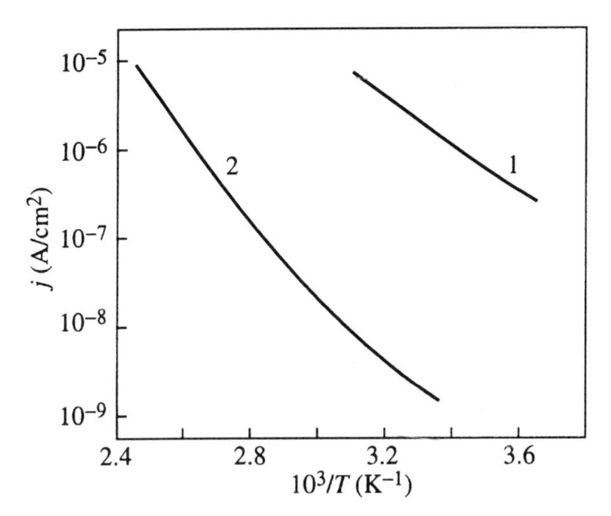

Figure 5.8. Temperature dependence of the volume component of a reverse current measured at 10 V.

10^{12} cm^{-3} within the temperature range 270–370 K depends exponentially on the inverse temperature (Fig. 5.8, curve 1). The ionization energy of the generation center, calculated from this dependency, is $E_V + 0.58$ eV, which corresponds to the ionization energy (determined from the capacitance measurements) of the center $E1 = E_C - 0.535$ eV. Since the product of the reverse current by the capacitance of the p-n junction does not depend on the applied voltage (Kostylev et al. 1989), the reverse current must be the generation-recombination current of Sah-Noyce-Shockley (Sah et al. 1957), and it is controlled by generation-recombination of the charge carriers through the $E1$ level in the space charge region.

The parameters of the centers responsible for the appearance of microplasmas were investigated in the p-n structures of two types (Vyzhigin et al. 1992). Interstitial defects $E1$ and $E4$ with concentration 10^{12} cm^{-3} dominated in the structures of the first type. Vacancy defects $E5$, whose concentration reached 2×10^{12} cm^{-3}, dominated in the structures of the second type. Curves 1 and 3 in Figure 5.9 show the temperature dependences of the first microplasma voltage for samples with DLCs. It is seen that these dependences are nonmonototic. This may be related to the recharge of

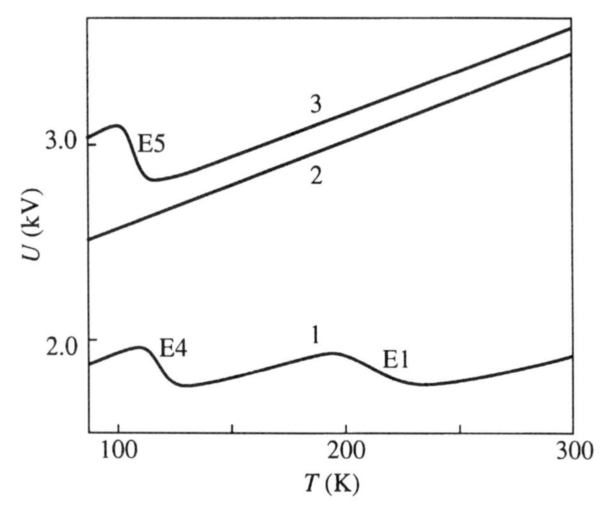

Figure 5.9. Temperature dependence of the voltage at which the first microplasma pulse appears. The appearance of microplasma is registrated 10 s after bias applied.

the donor DLCs in the space charge region of the microplasma channel. The emission rate of the carriers from DLCs into the conduction band at a fixed temperature was determined from the time dependence on the applied voltage of the appearance of the first microplasma pulse (Vyzhigin et al. 1988). The temperature dependencies of the rate of electron emission from the centers into the conduction band coincided with analogous dependencies measured for the centers $E1$, $E4$, and $E5$ by the DLTS technique. Thus it has been demonstrated that IPDs obtained under heat treatment participate in forming donor centers that are responsible for the appearance of microplasmas.

5.3. TECHNOLOGY OF MANUFACTURING HIGH-VOLTAGE POWER DEVICES

The above results showed that IPDs can play an essential part in forming electrically active centers. In order to preclude the formation of electrically active centers, or at least to diminish their concentration, it is necessary to prevent the generation of IPDs, or to provide the annihilation of IPDs of one type by introducing the defects of another type. Conducting heat treatment in the chlorine-containing atmosphere proved to be effective for controlling the processes of generation/suppression of IPDs (Sobolev and Shek 1979).

It was further shown by Lebedev et al. (1979) and Sobolev and Chelnokov (1987) that the annealing of high-voltage power structures with p-n junctions in a chlorine-containing atmosphere makes it possible to getter undesirable deep centers reducing the value of lifetime of minor carriers and causing the appearance of low-voltage microplasmas. Also the use of the chlorine-containing atmosphere allowed the doping impurity profile to be adjusted (Sobolev et al. 1984; Subrahmanyan et al. 1987).

A new technology of manufacturing high-voltage power devices has been developed. It is based on the role played by IPDs in forming electrically active centers. In this technology all high-temperature operations of oxidizing and diffusion of donor- and acceptor-doping impurities were conducted in a chlorine-containing atmosphere (Sobolev et al. 1984). This made it possible to manufacture a number of diodes and thyristors with operating voltages up to 10 kV and operating currents up to several kA.

Here we give some parameters of high-power high-voltage avalanche diodes, with the blocking voltage 4–6 kV and operating current 1250 A (Vil'yanov et al. 1989). The concentration of DLCs in the p^+-n-n^+ structures with an area of 23 cm^2 was 10^9–10^{10} cm^{-3}; the lifetime reached was 400 μs, and it provided an effective modulation of the n-base (the device thickness was 500–800 μm). The reverse current was made much lower: with the reverse voltage of the order of 100 V, the volume component of the reverse current at room temperature was $j_V \sim 10^{-9}$ A/cm^2. At $T > 60$°C, j_V did not depend on the applied voltage (Fig. 5.10) and was determined by the generation recombination of charge carriers in the neutral part of the n-base, that is, a Shockley diffusion current (Shockley 1950) with the activation energy equal to the energy gap of silicon (Fig. 5.8, curve 2). The experimental values of the diffusion current coincided well with the calculated values based on the lifetimes

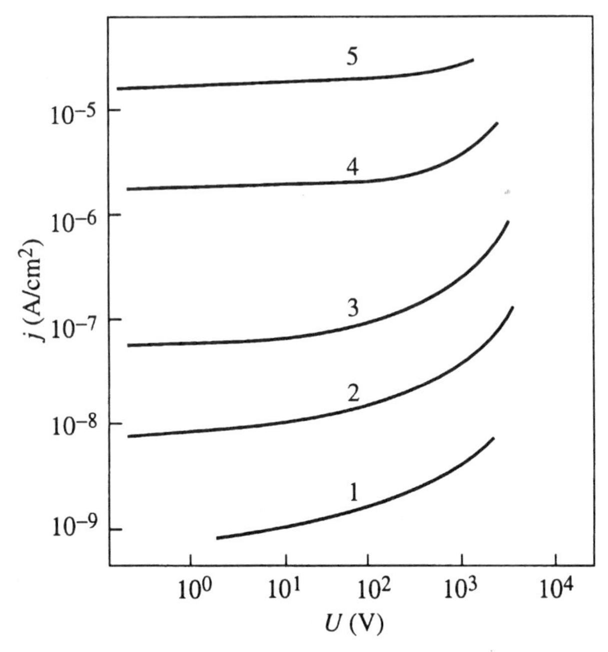

Figure 5.10. Voltage dependence of the bulk component of a reverse current measured at T (K): (1) 295, (2) 322, (3) 344, (4) 378, and (5) 408.

determined by the Lax technique. Previously the diffusion current in Si p-n junctions was observed only in low-voltage (10−15 V) structures with small areas (1−10 mm^2) and smaller volumes of the space charge region by four orders of magnitude than in the structures manufactured by us (Ivanov et al. 1982).

In the high-power high-voltage avalanche diodes, the breakdown voltage of the first microplasmas is only a few percent lower than the voltage at which the breakdown takes place over the entire area of the p-n junction. In these structures the voltage of the appearance of the first microplasma pulse increased monotonically with increasing temperature of measurement (Fig. 5.9, curve 2). The appearance of microplasmas in the structures was not caused by DLCs but by the nonuniformity of the resistivity distribution in silicon. The decrease of the concentration of electrically active centers that was achieved resulted in the decrease of the reverse current, at which the avalanche breakdown was reached practically all over the area of the diode structures by about an order of magnitude (to $1\,A/cm^2$).

5.4. CONCLUSION

We have demonstrated the significance of IPDs in defect formation in high-voltage power devices. The study of the behavior of swirl defects and of the diffusion of aluminium makes it possible to investigate processes of generation and relaxation of IPDs and determine their parameters.

The diffuse scattering of γ rays on the defects discovered here suggests that the formation of those defects is caused by the supersaturation of silicon with vacancies and self-interstitials. For the first time the generation of vacancy defects was observed directly after the heat treatment of Si in a chlorine-containing atmosphere.

We have also identified DLCs whose appearance is related to IPDs. The study of these DLCs makes it possible to investigate the effect of different technological factors on the processes of generation and relaxation of IPDs during the heat treatment of silicon.

We have seen that the annealing of radiation defects is accompanied by supersaturation of NTDS with self-interstitials and by the formation of structural defects and recombination centers of an interstitial type. To obtain NTDS of high quality (with a low concentration of structural defects and with long lifetime of the minority carriers), we have proposed that radiation defects be annealed in a chlorine-containing atmosphere.

Finally we have outlined the fundamentals of defect engineering in the technology of high-voltage power devices. The technology has been developed to fabricate such devices for operating currents of thousands of amperes and operating voltage up to several thousand volts. This technology is based on using the chlorine-containing atmosphere in the processes of oxidation and doping impurity diffusion.

The author wishes to thank Dr. N. M. Shmidt and Dr. O. V. Alexandrov for useful discussions of this chapter.

REFERENCES

Antognetty, P., Antoniadis, D. A., Dutton, R. W., and Oldham, W. G. (1983). *Process and Device Simulation for MOS-VLSI Circuits*. Martinus Nijhoff, The Hague.

Astrova, E. V., Voronkov, V. B., Kozlov, V. A., Lebedev, A. A., and Ecke, W. (1987). Preprint 1161 (in Russian). Ioffe Physico-Technical Institute, Academy of Sciences of the USSR, Leningrad.

Berman, L. S. and Lebedev, A. A. (1981). *Capacitance Spectroscopy of Deep Centers in Semiconductors*. Nauka, Leningrad (in Russian).

Caughey, D. M. and Tomas, R. T. (1967). *Proc. IEEE* 55, 2192–2193.

Chen, J.-W. and Milnes, A. G. (1980). *Ann. Rev. Mater. Sci.* 10, 157–228.

Crees, D. E. and Taylor, P. D. (1982). In *Proc. 4th Int. Conf. on Neutron Transmutation Doping of Semiconductor Materials*, Larrabee, R. D., ed., pp. 181–191.

d'Aragona, F. S. (1971). *Phys. Stat. Sol.* 7, 577–582.

de Kock, A. J. R. (1973). *Philips Res. Rep. Suppl.* 1, 1–102.

Fahey, P. M., Griffin, P. B., and Plummer, J. D. (1989). *Rev. Mod. Phys.* 61, 289–384.

Föll, H. and Kolbesen, B. O. (1975). *Appl. Phys.* 8, 319–331.

Gösele, U. and Tan, T. Y. (1983). *Mater. Res. Soc. Symp. Proc.* 14, 45–59.

Gossmann, H.-J., Gilmer, G. H., Rafferty, C. S., Unterwald, F. C., Boone, T., Poate, J. M., Luftman, H. S., and Frank, W. (1995). *J. Appl. Phys.* 77, 1948–1951.

Gresserov, B. N. and Sobolev, N. A. (1990). *Inorg. Mater.*, 26, 1503–1504.

Gresserov, B. N., Sobolev, N. A., Vyzhigin, Yu. V., Eliseev, V. V., and Likunova, V. M. (1991). *Sov. Phys. Semicond.* 25, 488–491.

Hu, S. M. (1974). *J. Appl. Phys.* 45, 1567–1573.

Ivanov, E. I., Lopatina, L. B., Sukhanov, V. L., Tuchkevich, V. V., and Shmidt, N. M. (1982). *Sov. Phys. Semicond.* 16, 129–132.

Jantsch, W., Wünstel, K., Kumagai, O., and Vogl, P. (1982). *Phys. Rev.* B25, 5515–5518.

Kimerling, L. C., Benton, J. L., and Rubin, J. J. (1981). *Inst. Phys. Conf. Ser.* 59, 217–222.

Kostylev, V. A., Nikolaev, Yu. A., Sobolev, N. A., Fedorov, L. M., and Shek, E. I. (1989). *Sov. Elect. Eng.* 60, 95–100.

Kurbakov, A. I., Rubinova, E. E., Sobolev, N. A., Trunov, V. A., and Shek, E. I. (1986). *Sov. Phys. Crystallogr.* 31, 582–586.

Kurbakov, A. I., Rubinova, E. E., Sobolev, N. A., Stuk, A. A., Trapeznikova, I. N., Trunov, V. A., and Shek, E. I. (1988). *Sov. Tech. Phys. Lett.* 14, 836–838.

Kurbakov, A. I. and Sobolev, N. A. (1994). *Mater. Sci. Eng.* B22, 149–158.

Lax, B. and Neustadter, S. T. (1954). *J. Appl. Phys.* 25, 1148–1154.

Lebedev, A. A., Sobolev, N. A., and Shek, E. I. (1979). *Electrotechnicheskaya Promyshlennost. Ser. Preobrazovatelnaya Technica* 2–4 (in Russian).

Meese, J. M., ed. (1979). *Proc. 2nd Int. Conf. on Neutron Transmutation Doping in Semiconductors*, Columbia, MO, April 23–26, 1978. Plenum Press, New York.

Mizuo, S. and Higuchi, H. (1982). *Jap. J. Appl. Phys.* 21, 56–60.

Paxman, D. H. and Whight, K. R. (1980). *Sol. St. Electr.* 23, 129–132.

Ravi, K. V. (1981). *Imperfection and Impurities in Semiconductor Silicon, Mobil Tyco Solar Energy Corporation, Waltham, Massachusetts.* Wiley-Interscience, New York.

Sah, C. T., Noyce, R. N., and Shockley, W. (1957). *Proc. IEEE* 45, 1228–1237.

Sah, C. T. and Wang, C. T. (1975). *J. Appl. Phys.* 46, 1767–1776.

Samara, G. A. and Barnes, C. E. (1987). *Phys. Rev.* B35, 7575–7584.

Senes, A. (1981). In *Proc. 3rd Int. Conf. on Neutron Transmutation Doping of Semiconductors*, Guldberg, ed., pp. 339–353.

Shiraki, H. (1976). *Jap. J. Appl. Phys.* 15, 1–10.

Shockley, W. (1950). *Electrons and Holes in Semiconductors.* Van Nostrand, New York.

Sobolev, N. A. and Shek, E. I. (1979). *Abstracts of 8th Vsesoyuznaya nauchno-technicheskaya conf. po problemam avtomatizirovannogo electroprivoda, silovykh poluprovodnikovykh priborov i preobrazovateley na ikh osnove*, October 16–19, 1979, Tashkent, USSR, pp. 37–38 (in Russian).

Sobolev, N. A., Chelnokov, V. E., and Shek, E. I. (1984). *Electrotechnicheskaya Promyshlennost. Ser. Preobrazovatelnaya Technica* 9, 15–17 (in Russian).

Sobolev, N. A., Shek, E. I., Dudavskii, S. I., and Kravtsov, A. A. (1985). *Sov. Phys. Tech. Phys.* 30, 842–843.

Sobolev, N. A. and Chelnokov, V. E. (1987). *Proc. 2nd Int. Autumn Meeting on Gettering and Defect Engineering in Semiconductor Technology*, Garzau, Germany, October 11–17, 1987, Richter, H., ed., pp. 179–184.

Sobolev, N. A., Vyzhigin, Yu. V., Eliseev, V. V., Kostylev, V. A., Likunova, V. M., and Sheck, E. I. (1989). *Diffus. Defect Data Sol. St. Data B. Sol. St. Phenomena* 6–7, 181–186.

Sobolev, N. A., Stuk, A. A., Kharchenko, V. A., Shek, E. I., and Minenko, S. V. (1990). *Inorg. Mater.* 26, 1342–1344.

Sobolev, N. A., Vyzhigin, Yu. V., Gresserov, B. N., Shek, E. I., Kurbakov, A. I., Rubinova, E. E., and Trunov, V. A. (1991). *Diffus. Defect Data Sol. St. Data B. Sol. St. Phenomena* 19–20, 169–174.

Sobolev, N. A., Kurbakov, A. I., Kyutt, R. N., Rubinova, E. E., Sokolov, A. E., and Shek, E. I. (1992). *Sov. Phys. Tv. Tela.* 34, 1365–1368.

Subrahmanyan, R., Massoud, H. Z., and Fair, R. B. (1987). *J. Appl. Phys.* 61, 4804–4807.

Taylor, W., Marioton, B. P. R., Tan, T. Y., and Gösele, U. (1989). *Radiat. Eff. Defects Sol.* 111–112, 131–150.

Vil'yanov, A. F., Vyzhigin, Yu. V., Gresserov, B. N., Eliseev, V. V., Likunova, V. M., Maksutova, S. A., and Sobolev, N. A. (1989). *Sov. Phys. Tech. Phys.* 34, 1184–1185.

Volle, V. M., Grekhov, I. V., Delimova, L. A., and Levinshtein, M. E. (1975). *Sov. Phys. Semicond.* 9, 429–432.

Voronov, I. N., Gres'kov, I. M., Grinshtein, P. M., Guchetl', R. I., Morokhovets, M. A., Sobolev, N. A., Stuk, A. A., Kharchenko, V. A., Chelnokov, V. E., and Shek, E. I. (1984). *Sov. Tech. Phys. Lett.* 10, 272–274.

Vysotskaya, V. V., Gorin, S. N., Gres'kov, I. M., Sobolev, N. A., Tkacheva, T. M., and Shek, E. I. (1988). *Inorg. Mater.* 24, 302–305.

Vyzhigin, Yu. V., Gresserov, B. N., and Sobolev, N. A. (1988). *Sov. Phys. Semicond.* 22, 330–332.

Vyzhigin, Yu. V., Zeman, J., Kostylev, V. A., Sobolev, N. A., and Shmid, V. (1989). *Sov. Phys. Semicond.* 23, 452–453.

Vyzhigin, Yu. V., Sobolev, N. A., Gresserov, B. N., and Sheck, E. I. (1991). *Sov. Phys. Semicond.* 25, 799–803.

Vyzhigin, Yu. V., Sobolev, N. A., Gresserov, B. N., and Shek, E. I. (1992). *Sov. Phys. Semicond.* 26, 1087–1091.

Yau, L. D. and Sah, C. T. (1974). *Sol. St. Electr.* 17, 193–201.

CHAPTER 6

ISOVALENT IMPURITY DOPING OF DIRECT-GAP III-V SEMICONDUCTOR LAYERS

V. V. CHALDYSHEV and S. V. NOVIKOV

The conventional way to control the properties of semiconductors consists in doping them with electrically active impurities, creating in the band gap shallow donor, shallow acceptor, or deep levels. In the case where the concentration of native defects is not high, this method allows one to change the material's conduction type and to control its resistivity in a wide range.

This chapter deals with controlling the properties of semiconductors in another way. This is isovalent (isoelectronic) impurity doping. The distinctive feature of that class of impurities is that they have a similar structure of exterior electronic shells and the same valences as the matrix atoms they are substituting. Compared to the electrically active impurities, the isovalent impurities introduce a relatively small perturbation into the electronic spectrum of a semiconductor. A typical concentration of isovalent impurities is 10^{18}–10^{20} cm^{-3}.

In wide- and indirect-gap III-V semiconductors, such as GaP, the perturbation in the electronic spectrum caused by the isovalent impurities, such as N or Bi, is sufficient for forming the local electron levels in the band gap. This leads to a number of well-known

Semiconductor Technology: Processing and Novel Fabrication Techniques,
Edited by M. Levinshtein and M. Shur.
ISBN 0-471-12792-2

effects, for instance, as described by Bergh and Dean (1976). We will not consider those effects in this chapter. We will consider another situation typical for direct-gap III-V semiconductors such as GaAs where the isovalent substitution does not result in the formation of local levels in the band gap.

Interest in the isovalent doping of direct-gap III-V semiconductors in the 1970s and 1980s appeared to be due to the possibility of growing bulk dislocation-free GaAs : In crystals by the Czochralski method. This attracted much attention and deserves special consideration (Winston et al. 1988). We will consider some relatively little-known phenomena related to the isovalent doping of direct-gap III-V semiconductor layers grown by different epitaxial methods or formed by ion implantation. We will show that in this case the isovalent doping results in a number of interesting effects caused by the interaction of isovalent impurities with the native lattice defects and with electrically active impurities. We will describe the most interesting phenomena and discuss the most important physical mechanisms related to isovalent impurity doping. We will use mostly GaAs as an example. We will skip all of the experimental details and describe some topics very briefly; additional information can be found in original papers listed in the References which is rather long but not exhaustive.

The chapter is organized in the following manner: In Section 6.1 we briefly discuss phenomena not directly linked to material growth technology. Among these are the variation of the band gap and of the lattice parameter due to isovalent impurity doping, the solubility limit of those impurities, and the alteration of the dislocation density in the epitaxial layers.

In Section 6.2 we consider the effects related to the isovalent doping of gallium arsenide with indium, antimony, and bismuth in liquid-phase epitaxy (LPE). We show that doping with indium and antimony leads to a considerable reduction in the concentration of deep centers typical for the LPE–GaAs. Doping with bismuth causes a considerable change in the incorporation of electrically active impurities into the growing layer. When undoped layers are grown, the concentration of electrically active background impurities strongly decreases, and the carrier mobility increases (the so-called purification effect).

In Section 6.3 we consider the isovalent impurity doping of GaAs in vapor-phase epitaxy (VPE) using either the chloride system or metalorganic compounds. We show that in VPE–GaAs : In such a doping causes a substantial reduction in the concentration of all deep levels, in particular, of the well-known EL2 centers. This effect, however, takes place only in a very narrow concentration range of indium.

In Section 6.4 we show that a substantial reduction of deep-level concentration is also observed in GaAs grown by molecular-beam epitaxy (MBE) and doped with either indium or antimony. In this case the changes are less drastic compared to the similar effects in VPE–GaAs : In, and they are closer to processes observed in LPE–GaAs : In and LPE–GaAs : Sb.

In Section 6.5 we consider the use of isovalent impurities in ion implantation. In this case the isovalent impurities enhance the activation of electrically active impurities and suppress radiation-induced defects.

In Section 6.6 we consider the peculiarities of isovalent doping of direct-gap III-V semiconductors such as InP, GaSb, InSb, and InAs. In particular, we show that by means of the isovalent impurity doping of LPE–GaSb, the concentrations of characteristic "native" acceptors can be considerably reduced.

In Section 6.7 we discuss the mechanisms responsible for the phenomena related to isovalent impurity doping of direct-gap III-V semiconductors. We show that irrespective of the compound, or kind of impurities, or method of material growth, the overwhelming majority of these phenomena can be understood in terms of three main mechanisms concerning the interaction of isovalent impurities with native point defects and electrically active impurities. Based on the understanding of these processes, one can control and optimize the material properties.

6.1. GENERAL PHENOMENA RELATED TO ISOVALENT IMPURITY DOPING

In this section we briefly discuss phenomena related to the isovalent impurity doping that are not related directly to any concrete materials growth technology. We consider band gap and lattice parameter variations and solubility limits of some isovalent impurities as well as the effect of isovalent impurities on the dislocation density and distribution in the epitaxial layers.

Indium and antimony are the typical isovalent impurities used in the epitaxial growth of GaAs. These impurities have an unlimited solubility in gallium arsenide. The interval of the composition of solid solutions containing the minor components of less than 1% (or $5 \times 10^{20}\ cm^{-3}$) is usually considered to be the region of isovalent doping.

As a result of the difference between the covalent radii of In (1.437 Å) and Sb (1.377 Å) compared to Ga (1.262 Å) and As (1.181 Å), respectively, the lattice parameter of the epitaxial layers of GaAs : In and GaAs : Sb is slightly greater than in the undoped GaAs. A small lattice mismatch at the interface GaAs/GaAs : In leads to the bending of dislocations, growing from the substrate, and to the decrease of the dislocation density in the epitaxial layers as shown in Figure 6.1 (Chaldyshev et al. 1995). Due to a similar structure of the external electronic shells for the

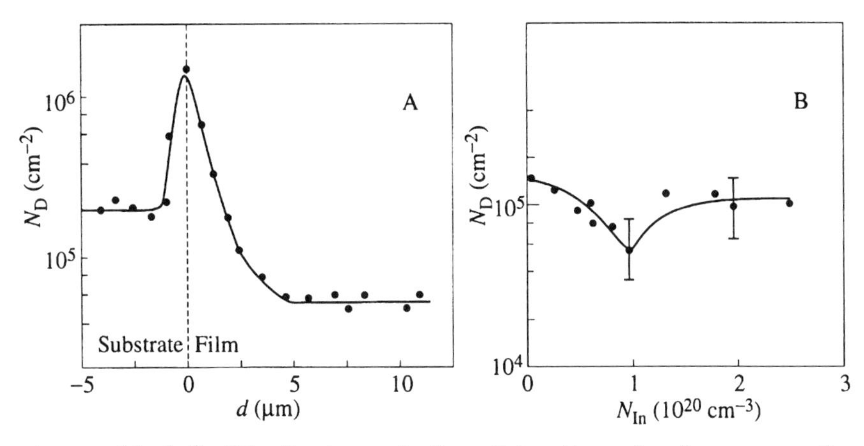

Figure 6.1. (A) Distribution of the dislocation density across the VPE–GaAs : In sample with the indium concentration of $1 \times 10^{20}\ cm^{-3}$; $d = 0$ corresponds to the substrate/film interface. (B) Dislocation density in the films (far from the substrate/film interface, $d > 5\ \mu m$ in panel A) as a function of the indium concentration.

isovalent pairs Ga-In and As-Sb, neither In nor Sb creates its own level in isovalent substitution. (Biryulin et al. 1981, 1983). However, heteroantisite defects (e.g., Sb_{Ga}) which are similar by their parameters to the native antisite defects (As_{Ga}) (Mitchel and Yu 1987; Yakimova et al. 1993; Samoilov et al. 1994) should be mentioned.

In or Sb doping leads to a decrease in the band gap of GaAs. This decrease described quantitatively by the well-known law for quasi-binary solid solutions $(AB)_x(CD)_{1-x}$, that is, $E_g(x) = E_g(0) - (E_g(0) - E_g(1))x - Cx(1-x)$, where $E_g(0)$, $E_g(1)$ are the band gaps of the binary components, and C is the empirical coefficient. For In and Sb in GaAs, C is equal to 0.4 eV and 1.2 eV, respectively. In the region of isovalent doping, namely for small concentrations of the minor component x, the dependence $E_g(x)$ is actually linear, and typical shifts in E_g do not exceed 20 meV.

Unlike In and Sb doping, bismuth doping of gallium arsenide causes no detectable changes either in the band gap or in the lattice parameter. This is apparently due to the low solubility of bismuth ($< 10^{18}\,cm^{-3}$) in GaAs at any temperatures typical for epitaxial growth (Akchurin et al. 1986a). In the photoluminescence spectra of the LPE–GaAs : Bi films, no lines that can be attributed to the isovalent center of Bi (Biryulin et al. 1987a) have been detected.

Band gap variations under the isovalent impurity doping have been studied for many III-V compounds and different isovalent impurities, such as GaSb : In (Biryulin et al. 1987), InP : Sb (Amus'ya et al. 1988), InP : Ga and InP : As (Pyshnaya et al. 1992), InSb : Bi (Lantsov et al. 1981), and GaSb : Bi (Germogenov et al. 1989).

No changes of E_g were detected in InP : Bi, perhaps due to the low solubility of bismuth, similar to that in GaAs (Akchurin et al. 1986a). Nevertheless, although the solubility of Bi in InSb and GaSb is limited, it can be as high as several percent and several tenths of a percent, respectively (Lantsov et al. 1981; Biryulin et al. 1988), which leads to a noticeable decrease of the band gap of the above-mentioned materials. The specific feature of these solid solutions is that the InBi compound is a semimetal with a tetragonal structure, while the GaBi compound is rather hypothetical. Another feature of bismuth doping of InSb and InSbAs solid solutions is a high concentration of Bi interstitials, which are shallow donors (Akchurin et al. 1982).

We have given here a brief review of the isovalent impurities in the III-V semiconductors. In the next sections we will consider the most important phenomena related to the interaction of isovalent impurities with lattice point defects and impurities. The experimental data described in Sections 6.2 to 6.6 will be summarized in Section 6.7.

6.2. LIQUID-PHASE EPITAXY

The pioneer investigations concerning the effects of isovalent impurities on the properties of the III-V epitaxial layers were done on gallium arsenide grown by liquid-phase epitaxy (LPE) and doped with indium and antimony (Solov'eva et al. 1981, 1982; Biryulin et al. 1981, 1983; Rytova et al. 1982; Mil'vidskii et al. 1983; see also the review by Bazhenov and Fistul' 1984). The phase diagrams of the Ga-In-As and Ga-As-Sb three-component systems have been well studied and were shown not to have any peculiarities for small concentrations of the minor component. The distribution coefficients of In and Sb at the temperatures typical for LPE GaAs are $2.4 \times 10^{-4} \exp(6575/T)$ for In and $1.27 \exp(4490/T)$ for Sb, where T is the temperature in K (Ganina et al. 1981).

One of the most interesting results obtained by isovalent indium and antimony impurity doping of GaAs epitaxial layers is the suppression of the deep levels related to the lattice intrinsic point defects and their complexes (this effect was discovered by Biryulin et al. 1983). Figure 6.2 shows a decrease of intensity of the photoluminescence line related to the deep center $E_v + 0.1$ eV, which is typical for LPE–GaAs. Later it was demonstrated (Kalukhov and Chikichev 1985; Mallik et al. 1994) that doping with indium and antimony also reduces the concentration of A and B hole traps (Fig. 6.3) typical for LPE–GaAs. This result is very important because these traps control the carrier lifetime in LPE–GaAs.

Figures 6.2 and 6.3 show that the effects of indium and antimony on the A, B and $E_v + 0.1$ eV centers are similar. This means that the mechanism of the effect is the same for a large number of different defects and does not depend on in which of the crystal sublattices the isovalent substitution is realized. This mechanism,

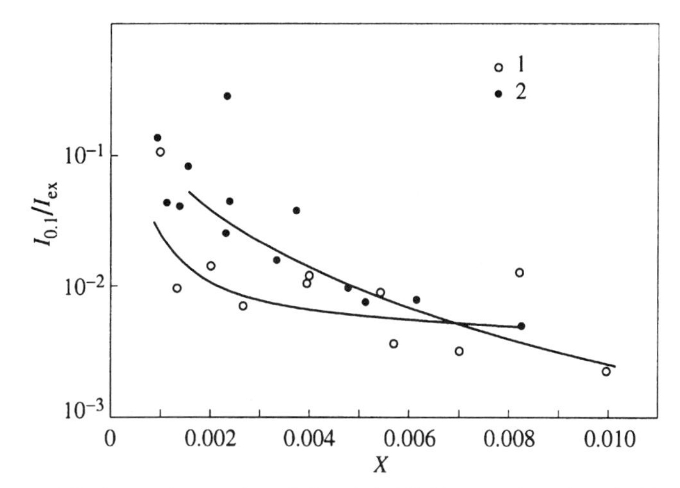

Figure 6.2. PL intensity ratio of the 1.4 eV line and excitonic line versus indium content in LPE–$Ga_{1-x}In_xAs$ (1) and antimony content in LPE–$GaAs_{1-x}Sb_x$ (2). The $I_{0.1}$ intensity is proportional to the $E_v + 0.1$ eV center concentration, $T = 2$ K.

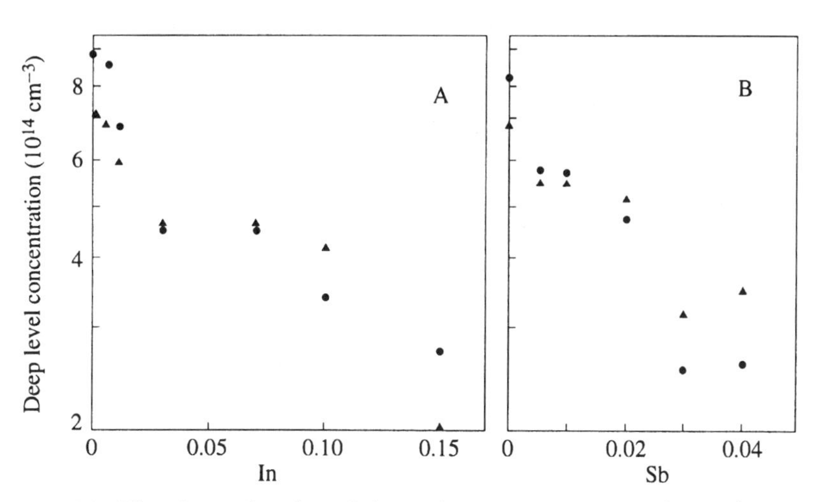

Figure 6.3. The dependencies of the *A* level concentration (circles) and *B* level concentration (triangles) in LPE–GaAs on In and Sb content (at. fraction) in the liquid phase (Kalukhov and Chikichev 1985).

which is based on the concept of local lattice strains caused by the In and Sb atoms, will be considered in Section 6.7.

Unlike the deep levels, indium and antimony doping of GaAs changes the concentration of shallow impurities slightly, usually within several tens of a percent. The amphoteric impurities of group IV (Si and Ge) are the most sensitive to the isovalent impurity doping. These impurities form the shallow donor levels substituting Ga and the shallow acceptor levels substituting As. It was found that the doping with indium facilitates the formation of the donor levels and that the doping with antimony facilitates the formation of the acceptor levels (Solov'eva et al. 1981, 1982). The redistribution of amphoteric impurities between sublattices due to isovalent impurity doping was detected by low-temperature photoluminescence (Biryulin et al. 1985). The main reason for this phenomenon seems to be a change in the component's liquid-phase activity when isovalent impurities are added. As a result the nonstoichiometry of the growing LPE–GaAs film and the probability of the impurity incorporation into the different sublattices change.

Let us consider another effect taking place in LPE–GaAs : In and in LPE–GaAs : Sb (Solov'eva and Mil'vidskii 1983). The dependencies of the electric properties on the isovalent impurity concentration showed that at a certain critical concentration of In or Sb, the electrical conductivity of the films abruptly decreases, compared either to undoped material or to material containing lower or higher concentrations of In and Sb. The critical concentration of In or Sb is $5–7 \times 10^{19}\ cm^{-3}$ and depends on the growth conditions. LPE–GaAs : In films were grown under identical technological conditions using pure gallium and pure bismuth as solvents (Vorob'eva et al. 1989). The critical concentration of indium was practically the same for both cases. This fact indicates that the effect is not related to the processes in the liquid phase and cannot be attributed to any specific defects and electrically active impurities. The nature of this phenomenon will be considered in Section 6.7.

Doping of GaAs with bismuth requires a special consideration. A low solubility of Bi in GaAs allows one to vary its content in the liquid phase within a very wide range up to the use of pure bismuth as a solvent. Though the solubilities of As in pure gallium and pure bismuth are practically equal, the dependence of the As solubility on the composition of the mixed Ga-Bi solvent (X_{Bi}) is not mono-

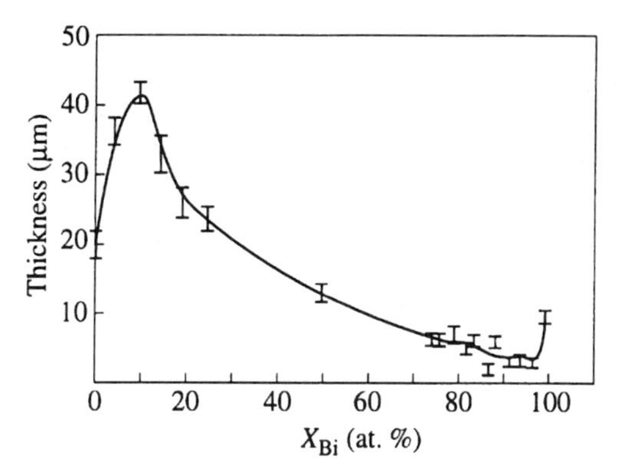

Figure 6.4. The thickness of LPE–GaAs films versus bismuth content X_{Bi} in the solvent. Growth time and temperature are the same for all the samples.

tonic (Yakusheva and Chikichev 1987). Figure 6.4 shows the thickness of the GaAs epitaxial layers grown by LPE under identical conditions as a function of the Ga-Bi solution composition. The solubility of As is maximal at X_{Bi} = 10–15 at.%, which results in a considerable increase of the growth rate (Biryulin et al. 1987b; Panek et al. 1986). With a further increase of X_{Bi}, the solubility of As and the thickness of the layers first decreases considerably, reaches the minimum at X_{Bi} = 90 at.%, and then starts rising again. An essential difference in the thickness of the layers grown from the pure gallium and pure bismuth solvents is caused partly by a notable difference in the diffusivity of arsenic in liquid Ga and liquid Bi and partly by a difference in the specific volume of Ga and Bi if the epitaxy is performed from the liquid phase of the fixed mass.

The investigation of the electrical properties of LPE–GaAs : Bi (Biryulin et al. 1987b) revealed only slight changes in the resistivity, concentration, and mobility of charge carriers when the bismuth content in the solvent varied from 0 to 70–80 at.% (Fig. 6.5). Comparing Figures 6.4 and 6.5, one can see that the increase in the growth rate and in the thickness of the layers in the region X_{Bi} = 10–15 at.% does not affect the electronic properties of the material. With $X_{Bi} > 80$ at.% the resistivity increases abruptly, reaching the maximum value at X_{Bi} = 90 at.% and then decreasing. As seen

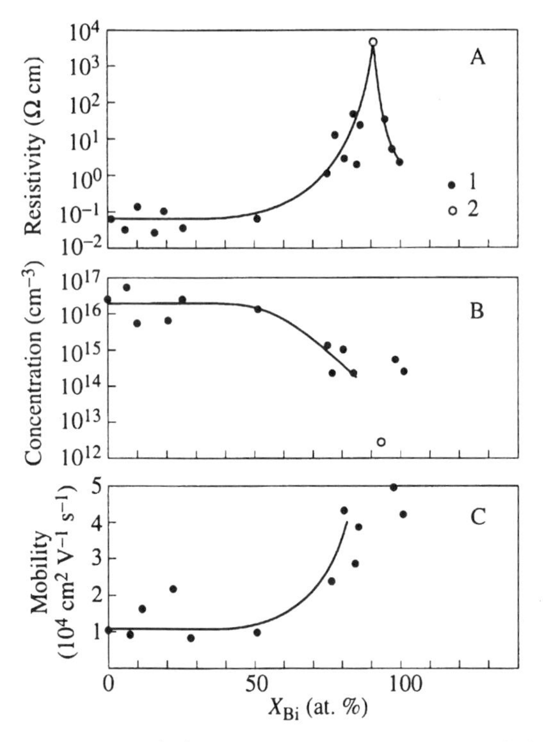

Figure 6.5. The resistivity (A) and carrier concentration (B) at 300 K and mobility (C) at 77 K in GaAs versus bismuth content in the solvent X_{Bi}. (1) n-type, (2) p-type (in resistivity maximum).

from Figure 6.5, the epitaxial films grown using a pure bismuth solvent ($X_{Bi} = 100$ at.%) have a considerably lower concentration and higher mobility of free electrons than the films obtained by the conventional LPE from pure gallium solution. (The technological conditions and the purity levels of the initial components were the same in both cases.) This phenomenon was discovered by (Ganina et al. 1982) and was called the "purification effect." The lowering of the growth temperature enhances the purification effect (Biryulin et al. 1986). Under the optimal conditions of the epitaxy from the bismuth solutions, it is possible to reproducibly obtain the GaAs layers with the electron mobility $\mu > 1 \times 10^5\ \mathrm{cm^2/V \cdot s}$ at 77 K (Biryulin et al. 1987c; Yakusheva et al. 1988, 1989) which is much higher than the values of μ typical for the layers grown under identical conditions from the gallium solutions.

From the analysis of the electrical parameters of the material, one can conclude that the purification effect is caused by a decrease in the concentration of background impurities. This conclusion was supported by PL studies (Biryulin et al. 1987a, 1987b, 1987c; Yakusheva et al. 1988, 1989), by Raman scattering experiments (Denisov et al. 1991), and by magnetospectroscopy measurements in the far IR region (Biryulin et al. 1987b; Yakusheva et al. 1988).

Sulfur is the most common background donor in the LPE–GaAs layers grown from the gallium solutions. Figure 6.6 shows the excitation spectra of shallow donors in the GaAs layers grown from the solutions with different bismuth content. One can see that the greater X_{Bi}, the smaller is the intensity of the peak related to the shallow S_{As} donor. As a result sulfur is no longer the main background donor in the layers obtained from the solutions with $X_{Bi} >$ 90 at.%. That causes a different behavior of the electrical parameters in this region (see Fig. 6.5).

A considerable change in the concentration was observed not only for sulfur but also for every other typical donor and acceptor background impurities, except possibly carbon. It was discovered that passing from the gallium to the bismuth solvent increases the

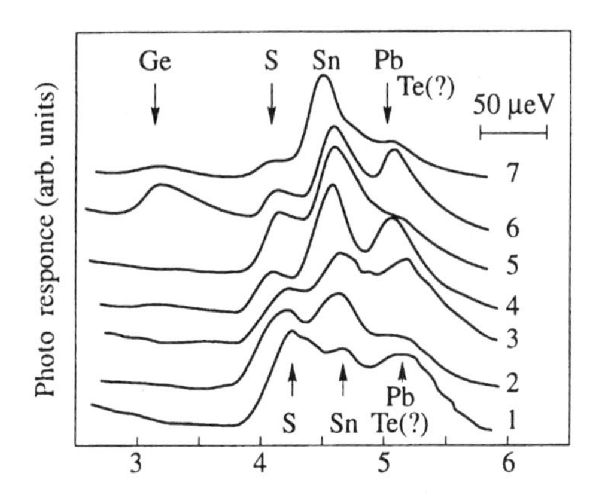

Figure 6.6. Photoexcitation spectra of shallow donors in gallium arsenide samples due to 1S-2P-1 transitions at 4.2 K in a magnetic field up to 6T (corresponding energy scale is presented in the insert). Concentration of bismuth in the solvent X_{Bi} (at.%): (1) 75; (2) 80; (3) 88; (4) 90; (5) 95; (6) 98; (7) 100.

concentration of impurities in gallium sublattice (the Zn, Be, Cu acceptors and the Si, Sn, Ge donors). In contrast, the concentration of impurities in arsenic sublattice decreases (the Ge, Si acceptors, the S, Te donors). Considerable changes take place in the native point defects as well. A clear manifestation of that phenomenon is the discovery in LPE–GaAs : Bi of a well-known EL2 deep level (Yakusheva et al. 1986; Brunkov et al. 1991). This level, which is attributed to the antisite defect As_{Ga}, is typical of gallium arsenide grown under stoichiometric or arsenic-rich conditions. It has never been observed in LPE–GaAs grown from the gallium solutions.

These facts indicate that the major reason for the purification effect and for the commensurable changes in the concentration of impurities and native defects is the change in the ratio of the main components' (As and Ga) concentrations in the liquid phase (by almost two orders) when bismuth substitutes gallium in the solvent. This leads to a decrease of the concentration of native defects and impurities in the anion sublattice and to an opposite effect in the cation sublattice of the crystal. Compared to the conventional gallium solvent, the use of bismuth yields the LPE of GaAs films under practically stoichiometric conditions, which allows one to obtain pure and perfect materials.

Switching from a gallium to a bismuth solvent, which leads to the corresponding changes in the activity of the components in the liquid phase, can considerably change the properties of the LPE–GaAs layers moderately and heavily doped with shallow impurities (Te, Sn, and Ge). It was found that increasing the bismuth content in the Ga-Bi solvent reduces the effective distribution coefficient of Te incorporated in the arsenic sublattice (Yakusheva and Sozinov 1986a; Yakusheva and Pogadaev 1991) and increases the effective distribution coefficient of tin, mostly incorporated in the gallium sublattice (Yakusheva and Beloborodov 1990). The Ge impurity required particular attention. When LPE is performed from gallium solutions, germanium is usually incorporated into the anion sublattice. This results in a p-type material with a low degree of compensation. In contrast, when GaAs is grown from a bismuth solvent, germanium is preferentially incorporated into the cation sublattice and the grown material is an n-type. It has a high concentration of electrons and a high degree of compensation. An inversion of the conductivity type takes place when the bismuth content in the solvent X_{Bi} exceeds 90 at.% (Yakusheva et al. 1985). The amphoteric properties of germanium can be described by a simple model that only considers the changes in component concentrations in the liquid Ga-As-Bi-Ge phase (Chaldyshev and Yakusheva 1989; Yakusheva and Pogadaev 1992). A high compensation degree of n-GaAs : Ge, Bi is the result of a very high concentration of deep levels (Biryulin 1987d; Chaldyshev and Yakusheva 1989a; Zhuravlev and Yakusheva 1990). A simple thermodynamic approach also allows one to describe the incorporation of Te and Sn impurities in the LPE of GaAs films from the Ga-Bi solutions (Yakusheva and Pogadaev 1992a).

6.3. VAPOR-PHASE EPITAXY

Isovalent impurity doping of GaAs in vapor-phase epitaxy (VPE) was realized using both the chloride system and metalorganic compounds. Two different versions of VPE gave similar results.

The doping with indium in the chloride system was realized by passing the flux of $AsCl_3 + H_2$ through an additional channel containing indium or by introducing the appropriate amount of indium into gallium. In the case of metalorganic VPE, trimethyl indium and trimethyl antimony were used as sources of In and Sb.

Indium doping caused a considerable reduction in the density of microdefects (Beneking et al. 1985) and a certain decrease in the dislocation density (Astrova et al. 1991). However, it only weakly affected the concentration and mobility of free electrons (Laurenti et al. 1989; Bykovskii et al. 1990; Astrova et al. 1991). It was found that the presence of indium prevented the chromium diffusion from the substrate into the epitaxial layer (Laurenti et al. 1989).

The most interesting effect that occurs during the isovalent indium doping of VPE–GaAs is a decrease of the deep center concentration. This phenomenon takes place when the layers are grown by VPE in a chloride system (Astrova et al. 1991; Chaldyshev et al. 1995) and by the metalorganic VPE (Beneking et al. 1985).

Figure 6.7 shows the spectra of deep-level transient spectroscopy (DLTS) of the VPE–GaAs : In layers (Astrova et al. 1991; Chaldyshev et al. 1995). The layers were grown in a $Ga/AsCl_3/H_2$

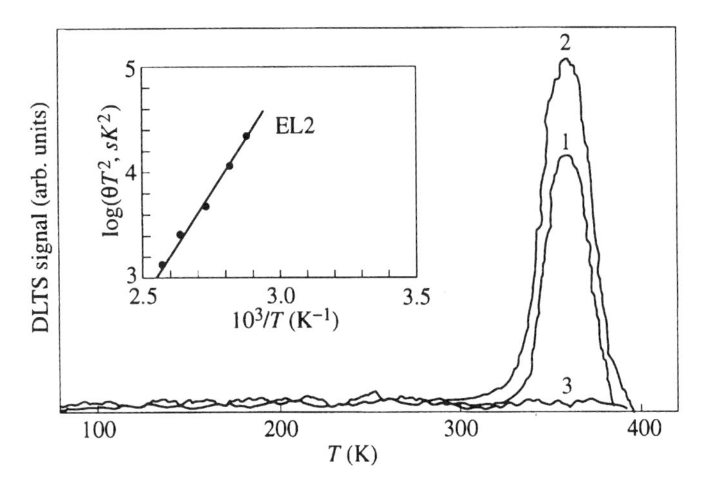

Figure 6.7. DLTS spectra of GaAs : S : In with the following indium content: Without In (1), $2 \cdot 10^{20}\ \mathrm{cm}^{-3}$ (2), $1 \cdot 10^{20}\ \mathrm{cm}^{-3}$ (3). The emission rate window at the peak maximum is 18.2 ms. An Arrhenius plot is shown in the insert that identifies the deep trap as EL2.

system and were additionally doped with sulfur ($n = (1\text{–}2) \times 10^{16}\ \mathrm{cm}^{-3}$) by introducing the mixture $SF_6 + He + H_2$ into the gas phase. In DLTS spectra of the indium-free samples, only one peak was observed to correspond to the electron trap in the middle of the band gap (Fig. 6.7, curve 1). Using Arrhenius plot (see insert in Fig. 6.7), this peak was identified as the well-known EL2 level. The concentration of EL2 level in the In-free layers was $(0.5\text{–}1.3) \times 10^{14}\ \mathrm{cm}^{-3}$. However, in the layer with indium concentration of $1 \times 10^{20}\ \mathrm{cm}^{-3}$ the concentration of EL2 level essentially decreased to a value less than the sensitivity threshold of the DLTS technique (see Fig. 6.7, curve 3). It should be noted that the effect of the substantial decreasing deep-level concentration was observed in a very narrow interval of indium concentration close to the $1 \times 10^{20}\ \mathrm{cm}^{-3}$ (Astrova et al. 1991; Chaldyshev et al. 1995). When indium concentration differs by a factor of two from the optimal value, the EL2-related DLTS peak appears, and the deep-level concentration is as high as that in the indium-free VPE–GaAs (see Fig. 6.7, curve 2). This effect seems to be important, since the EL2 traps control the carrier lifetime in a lightly doped VPE–GaAs.

A similar effect was found by PL study of VPE–GaAs heavily doped with sulfur ($n = (3–4) \times 10^{18}\,\text{cm}^{-3}$). In the PL spectra of this material, in addition to band-to-band (1.51 eV) and band-to-shallow-acceptor (1.49 eV) lines, deep-level related bands were observed at 0.9 and 1.2 eV. These lines are usually attributed to vacancy-donor complexes. Such defects are characteristic of heavily doped *n*-GaAs. Figure 6.8 shows the PL line intensities and the corresponding deep-center concentrations plotted with relation to the indium content. The deep-center concentrations were calculated using the data of the PL and Hall measurements (Astrova et al. 1991; Chaldyshev et al. 1995). The strong enhancement of the near-band-gap line intensity can be seen in Figure 6.8*A* at $N_{\text{In}} = 1 \times 10^{20}\,\text{cm}^{-3}$, which indicates an abrupt increase of the carrier lifetime. At the same time only a weak effect is observed for the deep-level related lines. This is due to a considerable reduction in the deep-level concentration (see Fig. 6.8*B*).

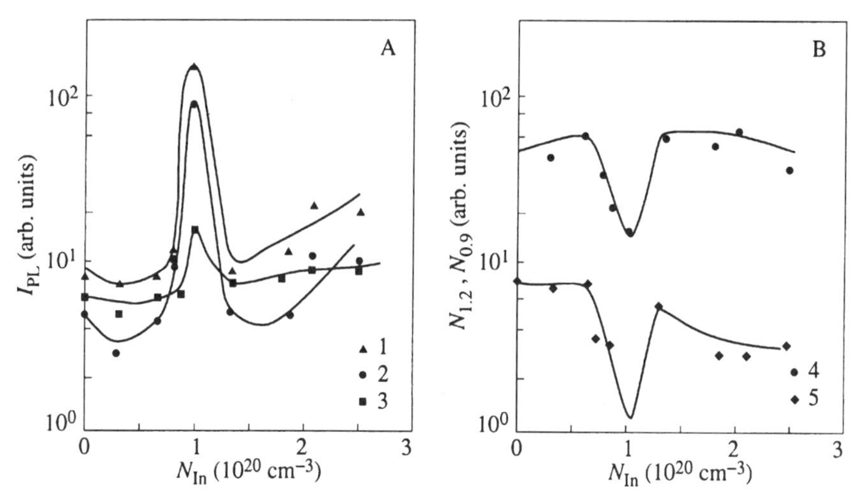

Figure 6.8. PL line intensities at 4.2 K (*A*) and relative deep-level concentrations (*B*) versus indium content in VPE-GaAs films. (*A*) Band-to-band luminescence (1), luminescence related to shallow donor–valence band transition (2), luminescence with photon energy ~ 1.2 eV (3). (*B*) Relative concentration of deep levels related to the lines at ~ 1.2 eV (4) and ~ 0.9 eV (5).

It should be noted again that the effect of the deep-level suppression is observed only in a very narrow range of indium concentrations. This feature seems to be the reason why this effect has not been observed in a number of papers (Laurenti et al. 1989; Coronado et al. 1986, 1987; Kol'chenko et al. 1989; Bykovskii et al. 1990, etc.).

6.4. MOLECULAR-BEAM EPITAXY

In and Sb are usually used as isovalent impurities in the molecular-beam epitaxy (MBE). Usually epitaxy is carried out under As-stabilized conditions. This corresponds to the (2 × 4)As surface reconstruction for (001) orientation of GaAs substrate, monitored by reflection high-energy electron diffraction (RHEED). Indium segregation on the growth surface is essential in the MBE of GaAs : In, especially at high substrate temperatures. This effect does not allow one to use Auger spectrometry for in-situ determination of the indium concentration in the bulk of the growing layer (Lubyshev et al. 1990). Indium segregation at the surface requires an As flux increase during the growth in order to maintain the As-stabilized surface structure near the transition from (2 × 4)As to (3 × 6). The sticking coefficients of In and Sb isovalent impurities quickly decrease with the increase in T_s from 500° to 600°C (Li et al. 1988). The indium doping results in a decrease of the dislocation density in the layer (Takeuchi et al. 1986) and in the formation of misfit dislocations at the substrate/film interface (Ioannou et al. 1988). Indium doping leads to a change in the shallow impurity concentration in the layers and, as a result, to a change in the concentration and mobility of charge carriers.

The GaAs : In films, doped with silicon in a wide range of concentrations, have been investigated by (Missous et al. 1987). It was found that the GaAs : In, Si layers were characterized by higher electron concentration and mobility compared to conventional GaAs : Si. This effect seems to be due to the indium influence on the distribution of the Si amphoteric impurity between the crystal sublattices and the resulting decrease of the self-compensation.

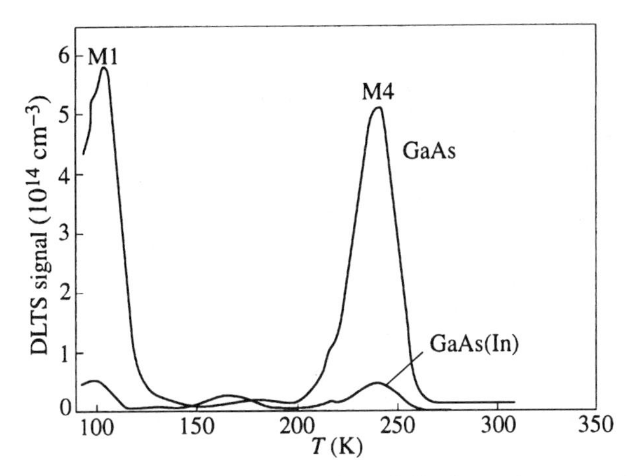

Figure 6.9. DLTS spectra for indium-doped GaAs and conventional MBE GaAs. The rate window is 200 s^{-1}. (Reprinted from Missous et al., 1987, Electrical properties of Indium doped GaAs layers grown by MBE, *J. Cryst. Growth* 81, 314–318, with kind permission of Elsevier Science–NL, Sara Burgerhartstraat 25, 1055 KV Amsterdam, The Netherlands.)

Figure 6.9 shows the DLTS spectra of the undoped and In-doped MBE-GaAs (Missous et al. 1987). In the DLTS spectrum of the undoped material, one can see the strong $M1$ and $M4$ peaks related to the deep levels typical for the MBE–GaAs. As can be seen from Figure 6.9, indium doping leads to a strong suppression of both deep levels. A similar effect was also observed for other deep levels (Bhattacharya et al. 1986). The suppression of deep levels was also detected by the PL study of the MBE–GaAs : In layers. A substantial decrease of the intensity of the peak related to the excitons bound on deep centers was observed (Lubyshev et al. 1990). This phenomenon was accompanied by the increase of total intensity of the radiative recombination (Lee et al. 1988). The suppression of deep levels was observed not only for indium doping but also for antimony doping (Li et al. 1988). In both cases the effect was very strong. Under optimal conditions the deep-level concentration can be reduced by two orders of magnitude. It should be mentioned that in the substrate temperature range from 530° to 575°C, the lower concentration of deep traps corresponded to higher T_s (Uddin and Anderson 1988).

6.5. ION IMPLANTATION

The main purpose of using the isovalent impurities in the ion implantation technology is to increase the activation efficiency of the electrically active impurities, such as Se or Si donors. This is very important at high implantation doses when the donor activation efficiency of the donors is rather low due to the formation of a large number of radiation defects and due to self-compensation during the postimplantation annealing. It seems that the additional dose of the implanted atoms should result in a greater concentration of radiation defects and, consequently, to a greater decrease of the donor activation efficiency. It is exactly what happens when in addition to Si or Se donors, the ions of inert gases, such as of Ar or Kr, are implanted (Abramov et al. 1991). However, the positive effect of the isovalent impurities on the activation of shallow donors during the postimplantation annealing in a number of cases overrides this obvious negative effect.

While In, Sb, and Bi are typical isovalent impurities in various epitaxial technologies of GaAs, in the case of ion implantation lighter elements such as P, N, and Ga are preferred, since they create a smaller number of radiation defects. Semi-insulating (100) GaAs substrates are normally used. Two alternative heat-treatment technologies were used for donor activation and annealing of the radiation-induced defects. The first technique is a rapid thermal annealing. The second method is thermal annealing in H_2 atmosphere under the protective cap of Si_3N_4.

The most studied pair of the donor and isovalent impurity used in ion implantation is Si + P (Hyuga et al. 1987; Morrow 1988; Abramov et al. 1991; etc.). It has been shown that Si + P co-implantation leads to the sheet electron concentration 1.5 times higher (Hyuga et al. 1987), or even 2 times higher (Abramov et al. 1991), than the corresponding value for GaAs implanted only with Si. Despite a higher electron concentration, the electron mobility in the layers co-implanted with Si + P was found to be higher than that in the layers implanted with Si alone.

Figure 6.10 shows the low-temperature PL spectra of the initial GaAs wafer (before the implantation) and of the samples implanted with Si and Si + P. The concentrations of the implanted silicon and

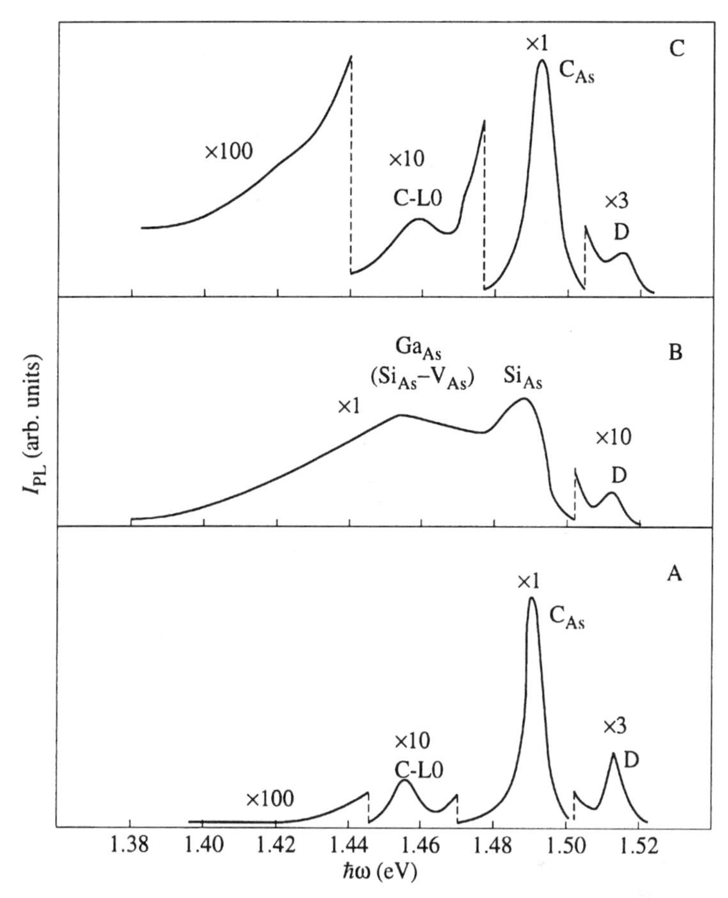

Figure 6.10. PL spectra of GaAs at 4.2 K. Annealing at 800°C (20 min). (*A*) GaAs wafer before implantation; (*B*) GaAs : Si; (*C*) GaAs : Si, P.

phosphorus were $1 \times 10^{18}\,\mathrm{cm}^{-3}$ and $5 \times 10^{18}\,\mathrm{cm}^{-3}$, respectively. The postimplantation annealing was carried out at 800°C for 20 minutes. The shallow-donor-related band *D* and the line related to the shallow background acceptor C_{As}, accompanied by the phonon replica, are the dominant radiative lines in the PL spectrum of the initial wafer (*A*). After the Si ion implantation two additional lines appeared in the PL spectrum (*B*). These are the lines related to Si_{As} shallow acceptor and 1.44 eV line related to silicon lattice point defect complex. These lines are typical for GaAs : Si when the implantation doses are high. As can be seen these lines are absent in the case of Si + P co-implantation (spectrum *C*). This means that the deep centers, responsible for the band 1.44 eV and the Si_{As} acceptors are not formed in any appreciable concentrations. Besides, the integral PL intensity in GaAs : Si, P is higher than in GaAs : Si. The analysis of the PL spectra shown in Figure 6.10 helps one to understand the causes of a higher donor activation efficiency in GaAs : Si, P compared to GaAs : Si. Phosphorus incorporation into the arsenic sublattice prevents silicon incorporation into this sublattice and thus facilitates the dominating formation of the Si_{Ga} donors. In other words, the implantation of phosphorus provides deviation from the stoichiometry, which suppresses the self-compensation of Si amphoteric impurity. In addition phosphorus prevents the diffusion of lattice defects and stimulates the mutual recombination of vacancies and interstitial atoms. This provides an additional improvement in the layer structure and properties.

Similar effects were observed for Se + Ga co-implantation. Since Se is a shallow donor when it substitutes As, Ga implantation provides an appropriate deviation from the stoichiometry and enhances the activation efficiency of donors. With the co-implantation of Se + Ga the sheet electron concentration in GaAs : Se, Ga was found to be 1.5 times higher than that in GaAs : Se (Abramov et al. 1991). Besides phosphorus, As and N ions were also used for co-implantation (Morrow 1988; Akimchenko et al. 1990). In both cases, however, no positive effect similar to that described above was achieved. Among other results on ion implantation of isovalent impurities in III-V compounds, we would like to mention the formation of p-type layers in GaAs : Ga (Tiginyanu et al. 1988) and the n-type layers in InP : P (Radautsan et al. 1988).

6.6. ISOVALENT IMPURITY DOPING OF OTHER DIRECT-GAP III-V COMPOUNDS (InP, GaSb, InSb, InAs)

Despite the considerable achievements in isovalent impurity doping of gallium arsenide, the investigation of corresponding effects in other direct-gap III-V compounds has attracted far less attention.

The studies of InP : Ga (Asahi et al. 1989; Pyshnaya et al. 1992), InP : As (Beneking et al. 1985; Westphalen et al. 1989; Pyshnaya et al. 1992), InP : Sb (Amus'ya et al. 1988; Bishop et al. 1988), and InP : Bi (Akchurin et al. 1986a) have revealed certain positive effects of isovalent impurities on the properties of indium phosphide. These effects were found to be similar to those in GaAs. However, the obtained experimental data were insufficient in order to understand an entire picture.

In narrow-gap III-V semiconductors, such as InAs, InSb, and their solid solutions, much attention was paid to the isovalent Bi impurity. The motivation for that was to the possibility of additional narrowing of E_g in InSbBi, InAsBi, and InAsSbBi solid solutions formed by substitution of the group V host atoms by bismuth (Lantsov 1981; Ma et al. 1989; Akchurin et al. 1992; Akchurin and Sakharova 1992a; Godaev et al. 1992). Besides, bismuth interstitials, forming the donor centers, are an important feature of these materials (Akchurin et al. 1982). The concentration of such donors can be rather high. An additional narrowing of E_g can be achieved using the effects of strong doping and compensation of Bi donors by shallow acceptors, such as zinc (Evgen'ev et al. 1985).

Isovalent impurity doping of GaSb was studied in the layers grown by liquid-phase epitaxy using indium and bismuth. The phase diagram of the Ga-Sb-In system is similar to that of the Ga-As-In system. According to the phase diagram, indium concentration in LPE−GaSb can vary in a wide range. However, the phase diagrams in the Ga-Sb-Bi and Ga-As-Bi systems are quite different. While the solubility of bismuth in GaAs is very small, in GaSb it can reach 0.3 at.%. The highest Bi concentration in the LPE−GaSb layers can be obtained not by growing from a pure bismuth solvent but rather by using a mixed Ga-Bi solvent (Biryulin et al. 1988). The doping of GaSb with bismuth leads to the formation of solid solutions with isovalent substitution and causes the band gap narrowing (Germogenov et al. 1989).

A specific feature of gallium antimonide is a high concentration of "native" acceptors. It varies from $5 \times 10^{16}\,\mathrm{cm}^{-3}$ in the epitaxial layers to $2 \times 10^{17}\,\mathrm{cm}^{-3}$ in the bulk crystals. The native acceptors are usually believed to be an isolated antisite defect $\mathrm{Ga_{Sb}}$ or a more complicated complex, which includes this defect. How to reduce the native acceptor concentration is the main problem to be solved for any growth method of GaSb. In liquid-phase epitaxy the concentration of native acceptors can be reduced by isovalent doping with indium (Biryulin et al. 1987). It was found that with the increasing In concentration in the layers, the concentration of the native acceptors is gradually decreasing, just like the concentration of centers $E_v + 0.1\,\text{eV}$ in the LPE–GaAs (see Fig. 6.3). At the optimal level of doping with indium (~ 0.1 at.%), the concentration of free holes at 300 K is a factor of two smaller than that of undoped GaSb grown under identical conditions. A stronger effect on the concentration of native acceptors in LPE–GaSb is obtained by the isovalent bismuth doping (Germogenov et al. 1990; Chaldyshev et al. 1992). Figure 6.11 shows the low-temperature PL spectra of undoped GaSb and GaSb doped with Bi. It shows free exciton (FE) and bound exciton ($BE1$–$BE4$) lines and also two lines (A and B) related to the native acceptors. It should be noted that the $BE1$ and $BE2$ peaks are also attributed to the recombination of excitons bound on the native acceptors. The PL spectra of Bi-doped samples are shifted toward greater wavelengths due to smaller energy gap of the $GaSb_{1-x}Bi_x$ solid solutions compared to GaSb (Germogenov et al. 1989). It should be pointed out that the intensities of the lines related to the native acceptors (A and B lines, $BE1$ and $BE2$ excitonic peaks) decrease strongly. This phenomenon is observed both in the layers grown from the Bi solutions (Zinov'ev et al. 1986; Germogenov et al. 1990) and from the mixed Ga-Bi solutions (Germogenov et al. 1990; Chaldyshev et al. 1992). Hall measurements in the GaSb : Bi films, made after the removal of the conductive substrate, showed that the free hole concentration was less than $1 \times 10^{16}\,\text{cm}^{-3}$. This result was also confirmed by the independent estimations of the free hole concentration using the differential thermo-emf technique. Hence the use of the isovalent Bi doping allows one to reduce the concentration of free holes and native acceptors in LPE–GaSb by an order of magnitude compared to the typical values for the epitaxial layers grown by the conventional LPE.

When LPE–GaSb films are doped with bismuth and tin, the results are similar to those described above for GaAs : Bi, Ge (Akchurin et al. 1992b). Just as Ge in GaAs, Sn in GaSb is an amphoteric impurity. In conventional LPE from the gallium solu-

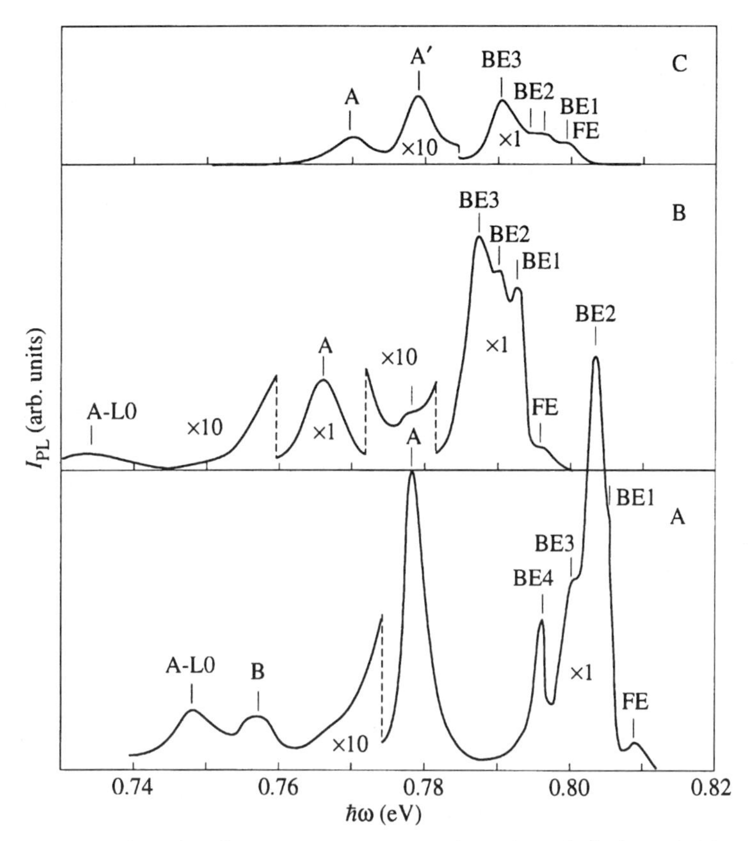

Figure 6.11. Photoluminescence spectra of undoped (A) and bismuth-doped (B, C) gallium antimonide at 4.2 K. Concentration of bismuth in the liquid phase (atomic fractions): (B) 0.12, (C) 0.84. Concentration in the solid phase (at.%): (B) 0.22; (C) 0.16.

tion, the Sn impurity preferentially forms a shallow acceptor center Sn_{Sb}. The material is p-type with a low degree of compensation. LPE from Ga-Bi solutions leads to the redistribution of tin between the crystal sublattices. With increasing the Bi content in the liquid phase, the degree of compensation of the GaSb : Bi, Sn layers also increases. When the bismuth content in the solution is about 90 at.%, the transition from p- to n-type conduction occurs. GaSb films grown from pure bismuth solutions are n-type.

6.7. CONCLUSION

The effects of isovalent impurity doping can be classified into three categories. The first group includes phenomena attributed to the deviation from stoichiometry. An example is the results of the co-implantation of electrically active and isovalent impurities. In this case the excess of group III or V components is formed in a direct way. Quite obvious are also the deviations from stoichiometry caused by the use of Ga-Bi and Bi solvents in the liquid-phase epitaxy. In this case one can observe strong changes in impurity incorporation into the different crystal sublattices, the amphoteric behavior of the group IV impurities, and the change of the nature of the dominant native point defects. Native point defect concentrations were calculated for GaAs, InAs (Akchurin et al. 1988), and GaSb (Akchurin 1993) doped with Bi. Shallow impurity incorporation and amphoteric impurity distribution between the sublattices were calculated for GaAs : Te, GaAs : Sn (Yakusheva and Pogadaev 1992a), GaAs : Ge (Chaldyshev and Yakusheva 1989; Yakusheva and Pogadaev 1992), and GaSb : Sn (Akchurin et al. 1992b) doped with Bi.

The effects related to the deviation from stoichiometry for isovalent indium and antimony doping in LPE are far less obvious and not so strong. For LPE–GaAs : In both calculations (Akchurin 1989) and experiments (Solov'eva et al. 1981; Biryulin et al. 1985) reveal the deviation of the composition of the grown material toward the increase of the concentration of the group V component. For LPE–GaAs : Sb the effect is the opposite. It is interesting that the deviations from stoichiometry in LPE–GaAs : In and LPE–GaAs : Sb are in the opposite direction compared to those in the ion implantation of the same impurities.

The change in deviation from stoichiometry during the molecular-beam epitaxy of GaAs : In is caused by indium segregation and by an increase of the arsenic flux needed in order to maintain the optimal surface reconstruction (Lubyshev et al. 1990). This effect leads to a redistribution of the Si amphoteric impurity between the sublattices and possibly plays a certain role in the deep-level suppression. Generally, despite a considerable difference in the growth conditions of MBE and LPE, these two methods are surprisingly alike with respect to isovalent impurity doping.

The second group includes the phenomena of healing lattice point defects by isovalent impurities. With the increase of the isovalent impurity concentration, this effect gradually intensifies until the densities of dislocations and the point defects begin to grow due to the substrate/film lattice mismatch. Examples of such a phenomenon are the decrease of concentrations of the center $E_v + 0.1\,\text{eV}$ and the hole traps in LPE–GaAs : In and LPE–GaAs : Sb, the reduction of deep-level concentration in MBE–GaAs : In and MBE–GaAs : Sb, the suppression of radiation defects in the GaAs : Si, P layers obtained by ion implantation, and the lowering of the native acceptor concentration in LPE–GaSb : In.

A possible mechanism responsible for these phenomena is the interaction of isovalent impurities with the lattice point defects through the local strain in the lattice. In the layers obtained by ion implantation, this interaction can lead to a recombination of the Frenkel pairs during the postimplantation annealing. In the epitaxial layers the local strain can cause an instability of various complexes of the lattice point defects and impurities. Due to the long range of the strain fields, this mechanism is independent of the sublattice in which the isovalent impurity is located. This point can be experimentally proved by comparing the data for GaAs : In and GaAs : Sb. For example, Figures 6.2 and 6.3 show effects that are similar for In and Sb doping and only slightly differ as a result of differences in atom sizes and phase diagrams.

A separate, third group includes phenomena that are realized in a very narrow range of isovalent impurity concentrations. These are an abrupt change in electrical properties of LPE–GaAs : In and LPE–GaAs : Sb, an abrupt reduction of the native acceptor concentration in LPE–GaSb : Bi, a change in electrical properties and a strong decrease of deep-level concentrations in VPE–GaAs : In. All these effects take place in a very narrow vicinity of a certain critical concentration of the isovalent impurity. For different materials and various growth conditions the critical concentration is in the range $4 \times 10^{19}\,\text{cm}^{-3}$ to $1 \times 10^{20}\,\text{cm}^{-3}$. It has been suggested (Solov'eva and Mil'vidskii 1983) that the critical concentration of the isovalent impurity corresponds to a situation when the local strains induced by the isovalent impurities in the crystal lattice overlap and cover the entire layer. In this model the critical concentration depends on the difference of covalent radii in the isovalent pairs Ga-In, As-Sb, and Sb-Bi. The growth technology can also influence the critical concentration due to a possible correlation in the spatial distribution of isovalent impurities (e.g., the formation of In-In pairs).

The general laws and the mechanisms described above allow us to predict the results of isovalent impurity doping before the growth experiments, taking into account the specific features of the method of the material growth. For instance, a specific feature of LPE with the Ga-Bi solutions is microinhomogeniety and separation of the liquid phase at low temperatures, which strongly affect the diffusion of the main components and impurities to the front of crystallization (Biryulin et al. 1987b).

In conclusion, in the modern growth technology isovalent impurity doping is a subtle instrument for controlling native point defects and electrically active impurities in III-V semiconductors. This approach finds wide application in the fabrication of semiconductor devices and allows us to optimize the properties of a material for specific practical tasks.

The authors are grateful to their teacher, late Professor Yu. V. Shmartsev. Under the guidance of Professor Shmartsev in late 1970s and early 1980s pioneering research of isovalent impurity doping of gallium arsenide were conducted at Ioffe Institute. The authors also appreciate the help of A. E. Kunitsyn in preparing the manuscript.

REFERENCES

Abramov, V. S., Akimchenko, I. P., Dravin, V. A., et al. (1991). *Sov. Phys. Semicond.* 25, 818–821.

Akchurin, R. Kh., Zinov'ev, V. G., Ufimtsev, V. B., et al. (1982). *Sov. Phys. Semicond.* 16, 126–129.

Akchurin, R. Kh., Le Din Kao, Nishanov, D. N., and Fistul' V. I. (1986). *Inorg. Mater.* 22, 5–8.

Akchurin, R. Kh., Biryulin, Yu. F., Islamov, S. A., Chaldyshev, V. V., and Shmartsev, Yu. V. (1986a). *Sov. Phys. Semicond.* 20, 794–796.

Akchurin, R. Kh., Donskaya, I. O., Dulin, S. I., and Ufimtsev, V. B. (1988). *Sov. Phys. Crystallogr.* 33, 464–470.

Akchurin, R. Kh. (1989). *Sov. Phys. Crystallogr.* 34, 520–523.

Akchurin, R. Kh., Sakharava, T. V., Tarasov, A. V., and Ufimtsev, V. B. (1992). *Inorg. Mater.* 28, 378–382.

Akchurin, R. Kh. and Sakharova, T. V. (1992a). *Sov. Tech. Phys. Lett.* 18, 16–20.

Akchurin, R. Kh., Zhegalin, V. A., and Chaldyshev, V. V. (1992b). *Sov. Phys. Semicond.* 26, 790–793.

Akchurin, R. Kh. (1993). *Sov. Phys. Crystallogr.* 38, 198–204.

Akimchenko, I. P., Dymova, N. N., Chaldyshev, V. V., and Shmartsev, Yu. V. (1990). *Sov. Phys. Semicond.* 24, 1155–1158.

Amus'ya, V. N., Biryulin, Yu. F., Vorob'eva, V. V., et al. (1988). *Sov. Phys. Semicond.* 22, 211–212.

Asahi, H., Sumida, H., Yu, S. J., et al. (1989). *Jap. J. Appl. Phys.* 28, L2119–L2121.

Astrova, E. V., Bobrovnikova, I. A., Vilisova, M. D., et al. (1991). *Sov. Phys. Semicond.* 25, 543–546.

Bazhenov, V. K. and Fistul', V. I. (1984). *Sov. Phys. Semicond.* 18, 843–853.

Beneking, H., Narozny, P., and Emeis, N. (1985). *Appl. Phys. Lett.* 47, 828–830.

Bergh, A. A. and Dean, P. J. (1976). *Light-Emitting Diodes*. Clarendon Press, Oxford.

Bhattacharya, P. B., Dhar, S., Berger, P., and Juang, F.-Y. (1986). *Appl. Phys. Lett.* 49, 470–472.

Biryulin, Yu. F., Ganina, N. V., and Chaldyshev, V. V. (1981). *Sov. Phys. Semicond.* 15, 1076–1078.

Biryulin, Yu. F., Ganina, N. V., Milvidskii, M. G., Chaldyshev, V. V., and Shmartsev, Yu. V. (1983). *Sov. Phys. Semicond.* 17, 68–71.

Biryulin, Yu. F., Ganina, N. V., Chaldyshev, V. V., and Shmartsev, Yu. V. (1985). *Sov. Phys. Semicond.* 19, 677–678.

Biryulin, Yu. F., Ganina, N. V., Chaldyshev, V. V., and Shmartsev, Yu. V. (1986). *Sov. Tech. Phys. Lett.* 12, 112–113.

Biryulin, Yu. F., Germogenov, V. P., Otman, Ya. I., et al. (1987). *Sov. Phys. Semicond.* 21, 681–685.

Biryulin, Yu. F., Golubev, L. V., Novikov, S. V., Chaldyshev, V. V., and Shmartsev, Yu. V. (1987a). *Sov. Phys. Semicond.* 21, 579–580.

Biryulin, Yu. F., Vorob'eva, V. V., Golubev, L. V., et al. (1987b). *Sov. Phys. Semicond.* 21, 579–580.

Biryulin, Yu. F., Nikitin, V. G., Nugmanov, D. L., and Chaldyshev, V. V. (1987c). *Sov. Tech. Phys. Lett.* 13, 527–528.

Biryulin, Yu. F., Vorob'eva, V. V., Golubev, L. V., et al. (1987d). *Sov. Tech. Phys. Lett.* 13, 530–531.

Biryulin, Yu. F., Germogenov, V. P., Ivleva, O. M., et al. (1988). *Sov. Tech. Phys. Lett.* 14, 719–720.

Bishop, S. G., Shanabrook, B. V., Klein, P. B., and Henry, R. L. (1988). *Phys. Rev. B* 38, 8469–8472.

Brunkov, P. N., Gaibullaev, S., Konnikov, S. G. et al. (1991). *Sov. Phys. Semicond.* 25, 205–207.

Bykovskii, V. A., Ivanyutin, L. A., Kol'chenko, T. I., et al. (1990). *Sov. Phys. Semicond.* 24, 46–49.

Chaldyshev, V. V., Astrova, E. V., Lebedev, A. A., et al. (1995). *J. Cryst. Growth* 146, 246–250.

Chaldyshev, V. V. and Yakusheva, N. A. (1989). *Sov. Phys. Semicond.* 23, 26–28.

Chaldyshev, V. V. and Yakusheva, N. A. (1989a). *Sov. Phys. Semicond.* 23, 137–138.

Chaldyshev, V. V., Germogenov, V. P., and Shmartsev, Yu. V. (1992). *Key Elect. Mater.*, 65, 109–116.

Coronado, L. M., Abril, E. J., and Aguilar, M. (1986). *Jap. J. Appl. Phys.* 25, L899–L901.

Coronado, L. M., Abril, E. J., and Aguilar, M. (1987). *Jap. J. Appl. Phys.* 26, L193–L195.

Denisov, V. N., Mavrin, B. N., Novikov, S. V., Chaldyshev, V. V., and Shmartsev, Yu. V. (1991). *Sov. Phys. Semicond.* 25, 898–900.

Evgen'ev, S. B., Kononkova, N. N., Lapkina, I. A., Sorokina, O. V., and Ufimtsev, V. B. (1985). *Inorg. Mater.* 21, 417–419.

Ganina, N. V., Mil'vidskii, M. G., and Ukhorskaya, T. A. (1981). *Inorg. Mater.* 17, 1138–1140.

Ganina, N. V., Ufimtsev, V. B., and Fistul, V. I. (1982). *Sov. Tech. Phys. Lett.* 8, 620–623.

Germogenov, V. P., Otman, Ya. I., Chaldyshev, V. V., and Shmartsev, Yu. V. (1989). *Sov. Phys. Semicond.* 23, 942–943.

Germogenov, V. P., Otman, Ya. I., Chaldyshev, V. V., Shmartsev, Yu. V., and Epiktetova, L. E. (1990). *Sov. Phys. Semicond.* 24, 689–693.

Godaev, O. A., Georgitse, E. I., Gutsulyak, L. M., et al. (1992). *Sov. Phys. J.* 35, 126–127.

Hyuga, F., Yamazaki, H., Watanabe, K., and Osaka, J. (1987). *Appl. Phys. Lett.* 50, 1592–1594.

Ioannou, D. E., Huang, Y. J., and Iliadis, A. A. (1988). *Appl. Phys. Lett.* 52, 2258–2260.

Kalukhov, V. A. and Chikichev, S. I. (1985). *Phys. Stat. Sol. (a)* 88, K59–K61.

Kol'chenko, T. I., Lomako, V. M., Rodionov, A. V., and Sveshnikov, Yu. N. (1989). *Sov. Phys. Semicond.* 23, 391–393.

Lantsov, A. F., Akchurin, R. Kh., and Zinov'ev, V. G., (1981). *Inorg. Mater.* 17, 1146–1148.

Laurenti, J. P., Wolter, K., Roentgen, P., Seibert, K., and Kurz, H. (1989). *Phys. Rev. B* 39, 5934–5946.

Lee, M. K., Chiu, T. H., Dayem, A., and Agyekum, E. (1988). *Appl. Phys. Lett.* 53, 2653–2655.

Li, A. Z., Kim, J. C., Jeong, J. C., et al. (1988). *J. Appl. Phys.* 64, 3497–3504.

Lubyshev, D. I., Migal', V. P., Preobrazhenskii, V. V., et al. (1990). *Sov. Phys. Semicond.* 24, 1158–1161.

Ma, K. Y., Fang, Z. M., Jaw, D. H., et al. (1989). *Appl. Phys. Lett.* 55, 2420–2422.

Mallik, K., Dhar, S., and Sinha, S. (1994). *Semicond. Sci. Technol.* 9, 1649–1653.

Missous, M., Singer, K. E., and Nicholas, D. J. (1987). *J. Cryst. Growth* 81, 314–318.

Mitchel, W. C. and Yu, P. W. (1987). *J. Appl. Phys.* 62, 4781–4785.

Morrow, R. A. (1988). *J. Appl. Phys.* 64, 1889–1896.

Panek, M., Ratuszek, M., and Tlaszala, M. (1986). *J. Mater. Sci.* 21, 3977–3980.

Pyshnaya, N. B., Radautsan, S. I., Chaldyshev, V. V., Chumak, V. A., and Shmartsev, Yu. V. (1992). *Sov. Phys. Semicond.* 26, 972–974.

Radautsan, S. I., Tiginyanu, I. M., and Pyshnaya, N. B. (1988). *Phys. Stat. Sol. (a)* 108, K59–K61.

Rytova, N. S., Solov'eva, E. V., and Mil'vidskii, M. G. (1982). *Sov. Phys. Semicond.* 16, 951–953.

Samoilov, V. A., Yakusheva, N. A., and Prints, V. Y. (1994). *Semicond.* 28, 901–905.

Solov'eva, E. V., Rytova, N. S., Mil'vidskii, M. G., and Ganina, N. V. (1981). *Sov. Phys. Semicond.* 15, 1243–1246.

Solov'eva, E. V., Mil'vidskii, M. G., and Ganina, N. V. (1982). *Sov. Phys. Semicond.* 16, 1161–1165.

Solov'eva, E. V. and Mil'vidskii, M. G. (1983). *Sov. Phys. Semicond.* 17, 1289–1291.

Takeuchi, H., Shinohara, M., and Oe, K. (1986). *Jap. J. Appl. Phys.* 25, L303–L305.

Tiginyanu, I. M., Pyshnaya, N. B., Spitsyn, A. V., and Ursaki, V. V. (1988). *Sov. Phys. Semicond.* 22, 1147–1149.

Uddin, A. and Anderson, T. G. (1988). *J. Appl. Phys.* 65, 3101–3106.

Vorob'eva, V. V., Zushinskaya, O. V., Novikov, S. V., Savel'ev, I. G., and Chaldyshev, V. V. (1989). *Sov. Phys. Tech. Phys.* 59, 951–953.

Westphalen, R., Jurgensen, H., and Balk, P. (1989). *J. Cryst. Growth* 96, 982–984.

Winston, H., Hunter, A. T., Kimura, H., and Lee, R. E. (1988). In *Semiconductors and Semimetals*. Willardson, R. K., and Beer, A. C., eds., vol. 26, pp. 99–141. Academic Press, San Diego.

Yakimova, R., Pashkova, T., and Hardalov, Ch. (1993). *J. Appl. Phys.* 74, 6170–6173.

Yakusheva, N. A., Sikorskaya, G. V., and Sozinov, V. N. (1985). *Inorg. Mater.* 21, 458–460.

Yakusheva, N. A., Prinz, V. Yu., and Bolkhovityanov, Yu. B. (1986). *Phys. St. Sol. (a)* 95, K43–K46.

Yakusheva, N. A. and Sozinov, V. N. (1986a). *Inorg. Mater.* 22, 475–478.

Yakusheva, N. A. and Chikichev, S. I. (1987). *Inorg. Mater.* 23, 1418–1421.

Yakusheva, N. A., Zhuravlev, K. S., and Shegai, O. A. (1988). *Sov. Phys. Semicond.* 22, 1320–1322.

Yakusheva, N. A., Zhuravlev, K. S., Chikichev, S. I., and Shegai, O. A. (1989). *Cryst. Res. Technol.* 24, 235–246.

Yakusheva, N. A. and Beloborodov, G. M. (1990). *Inorg. Mater.* 26, 4–7.

Yakusheva, N. A. and Pogadaev, V. G. (1991). *Inorg. Mater.* 27, 950–954.

Yakusheva, N. A. and Pogadaev, V. G. (1992). *J. Cryst. Growth* 123, 479–486.

Yakusheva, N. A. and Pogadaev, V. G. (1992a). *Cryst. Res. Technol.* 27, 21–30.

Zhuravlev, K. S. and Yakusheva, N. A. (1990). *Sov. Phys. Semicond.* 24, 523–527.

Zinov'ev, V. G., Morgun, A. I., Ufimtsev, V. B., and Arshavskii, A. N. (1986). *Sov. Phys. Semicond.* 20, 209–210.

CHAPTER 7

SURFACE PASSIVATION OF III-V COMPOUNDS BY INORGANIC DIELECTRICS AND POLYIMIDES

A. T. GORELENOK, N. D. ILYINSKAYA, I. A. MOKINA,
and N. M. SHMIDT

Passivation of semiconductor surfaces is performed in order to minimize the effect of the surface electronic processes on the bulk characteristics of semiconductor. Passivation also minimizes the variations of the bulk characteristics with time caused by various external effects (physicochemical, electric, or thermal). Passivation controls such parameters as the density of surface states, the surface recombination velocity, the flat band voltage, and the leakage current of reverse biased p-n junctions.

In this chapter we will consider two promising methods of passivating III-V compounds: passivation by inorganic dielectrics and passivation of InP, InGaAs, and InGaAsP by polyimide.

For the passivation of devices (mainly photodetectors) based on InP and InGaAsP, inorganic dielectrics such as SiO_2, Si_3N_4, Al_2O_3, and SiN_x are used (Gardner et al. 1984, Johnson and Kapoor 1991). Attempts have been made to use organic dielectrics for passivating InGaAs p-n junctions (Yeats and Dessonneck 1984). However, organic dielectrics have not been used so far in the mainstream technology of III-V compound devices. Nevertheless, these materi-

Semiconductor Technology: Processing and Novel Fabrication Techniques,
Edited by M. Levinshtein and M. Shur.
ISBN 0-471-12792-2

als are used successfully in silicon technology (Soone and Martynenko 1989), and they are of a great practical interest due to their simplicity and low cost.

In terms of surface passivation for the devices based on InP and InGaAsP, we will distinguish two types of devices.

Less strict requirements have to be met for the passivation of a surface that is not an active element of the device, such as the periphery of a p-n junction or Schottky barrier diodes, especially if the values of the volume component of the leakage current are large enough (more than several nA). We will refer to such devices as to the type A devices.

Much more rigid requirements are placed to the passivation of devices whose surface is a part of an active device area, for example, in field effect transistors with an insulated gate and in MIS structures. We will call such devices type B devices.

7.1. PASSIVATION OF SURFACES FOR TYPE A DEVICES

For type A devices based on InP and InGaAsP, surfaces have been successfully passivated by inorganic dielectrics. Using inorganic dielectrics, different research groups have managed to reduce the density of surface states from 10^{13} $eV^{-1}cm^{-1}$ to $(1-5) \times 10^{11}$ eV^{-1} cm^{-1} (Gardner et al. 1984; Belyakova et al. 1989; Johnson and Kapoor 1990; Hollinger et al. 1990). The surface recombination velocity was reduced down to $\sim 10^3$ $cm \cdot s^{-1}$ (Gorelenok et al. 1985a; Belousov et al. 1990).

These results made it possible to produce photodetectors based on InP/InGaAs (InGaAsP) structures operating at current densities of 1–40 $\mu A \cdot cm^{-1}$ (Gorelenok et al. 1985b; Chané et al. 1986; Bauer and Trommer 1988; Le Belligo et al. 1990) and even at the current density of 0.01 $\mu A \cdot cm^{-1}$ (Gorelenok et al. 1985a).

There are also certain achievements regarding the long-term stability of electric parameters of type A devices, even though the data given in the publications are scarce.

Stable values of reverse leakage current $I \leq 1$ nA were observed during 10^3 hours at 200°C for InGaAs/InP p-i-n photodiode passivated with SiN_x (Bauer and Trommer 1988). The reverse leakage current $I \leq 2$ nA was achieved (Chané et al. 1986) for the InGaAs/InP p-i-n photodiode at 10 V bias after 5000 hours of operation at +65°C. The passivating dielectric was SiO_2.

The process of passivation with *organic dielectrics* is much simpler and cheaper than that with inorganic dielectrics. Besides, it yields a low density of defects, allows surface planarization, and simplifies photolithography (in the case of light-sensitive polyimides) (Goncharova 1989).

Polyimides have long been used in the technology of silicon devices (Soone and Martynenko 1989). However, the data regarding the application of polyimides in the technology of devices based on InP/InGaAsP are rather limited. Perhaps only the publication (Yeats and Von Dessonneck 1984) should be mentioned. It shows the possibility of passivation of p-n junctions based on InGaAs/InP with polyimide. But in this work the effect of passivation was not studied in detail. In particular, the time stability of the dark leakage current was not checked.

Our investigations have shown that polyimide can be used quite successfully for the passivation of the devices based on InP and InGaAsP.

We investigated passivation with polyimide on the InGaAsP/InGaAs/InP mesastructures shown schematically in Figure 7.1. The method of obtaining these structures and their parameters are described by Gorelenok et al. (1985b). These structures provide a basis for the fabrication of the photodetectors for the spectral region of 0.8–1.67 μm. The polyimide AD9103 with resistivity of

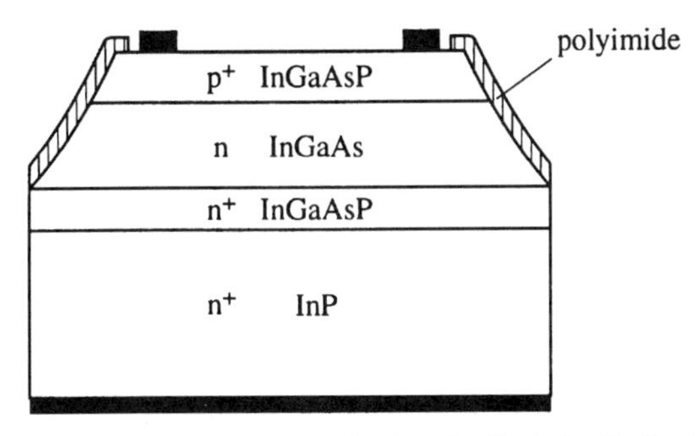

Figure 7.1. Cross-sectional view of the InGaAsP/InGaAs/InP mesastructure.

10^{16} $\Omega \cdot$ cm, and the coefficient of thermal expansion $\sim (3\text{–}4) \times 10^{-5}\,°C^{-1}$), was spun on by centrifuging, leaving a film that was 1–1.5 μm thick. While drying, the spun-on film was polymerized, so drying was an essential operation. Also the kinetics of the polymerization determines both the process of forming a spatial structure of polyimide and its dielectric and mechanical properties.

It was shown (Biernath and Soane 1989) that the compressive stresses that appear in the polymer film during polymerization depend nonlinearly on the temperature of polymerization and on the rates of heating and cooling. For fast heating and cooling (10°C $\cdot$ min^{-1}), the compressive stress is three times greater than that for the slow process (1°C $\cdot$ min^{-1}). These regimes were chosen as the limiting cases.

The efficiency of the passivation was estimated from the comparative measurements of the leakage current before and after the passivation. The measurements were performed on p-n junctions with a diameter of 100 μm.

The dashed line 1 in Figure 7.2 shows the current-voltage (I-V) characteristic of one of the mesastructures with a leakage current close to the minimal observed currents. Curve 1 was obtained for the as-etching surface without passivation.

Curves 2 to 4 show the I-V characteristics for the same sample with the polyimide coating obtained at different regimes of thermal polymerization (heating and cooling). Curves 3 and 4 were obtained using fast and slow heating and cooling regimes (10°C $\cdot$ min^{-1} and 1°C $\cdot$ min^{-1}, respectively). In both cases there was a considerable increase of the current compared to curve 1. The measurements for p-n junctions with diameters from 50 to 500 μm revealed that the current increase is related to an abrupt increase of the surface current and that the surface recombination velocity increases greatly, especially for reverse biases $U > 5$ V (Fig. 7.2).

However, it was possible to choose a regime of polymerization that allowed us to retain the minimal values of leakage currents (Fig. 7.2, curve 2). This regime was characterized by the heating and cooling of the polyimide from room temperature to 350°C at the rate of 2°C $\cdot$ min^{-1}.

Moreover passivation in the above optimal regime for part of the mesastructures could decrease the leakage currents by several orders of magnitude compared to the initial values. Curve 5 in Fig-

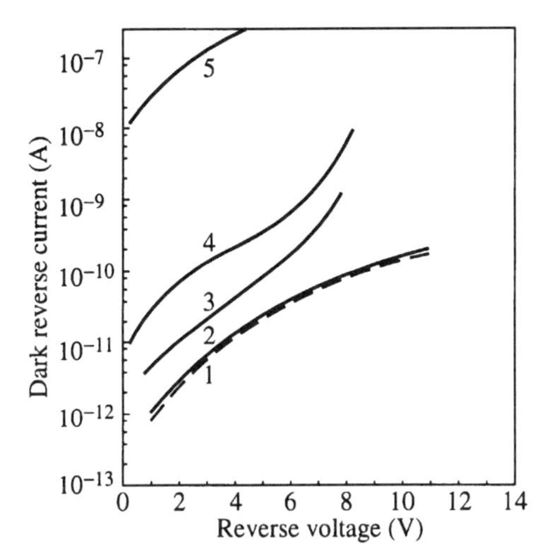

Figure 7.2. Reverse current–voltage characteristics of mesastructures fabricated on the same wafer at 300 K. *I-V* characteristics of different mesastructures after the etching (1, 5); *I-V* characteristics of mesastructures after the polyimide passivation under different regimes (2–4): (2) the optimal regime; (3) fast (10°C · min^{-1}) heating and cooling; (4) slow (1°C · min^{-1}) heating and cooling.

ure 7.2 shows the current-voltage characteristic of one of the samples immediately after etching but before the passivating coating was spun on. It is seen that the leakage current is four orders of magnitude greater than that for the best devices (compare to curve 1).

After applying a passivating polyimide coating at the above-mentioned optimal regime, the *I-V* characteristic of this sample (curve 2) practically coincided with curve 1 in Figure 7.2. It is important to note that this effect was reproducible.

Figure 7.3 shows a histogram of the leakage current distribution at the reverse bias of 5 V for the mesastructures before (shaded region) and after (unshaded region) the passivation. As seen in the figure, after passivation the percentage of mesastructures with leakage currents less than 10^{-9} became twice as large, while the percentage of devices with very large values of leakage current (10^{-8}–10^{-7} A) became three times as small. We believe that the

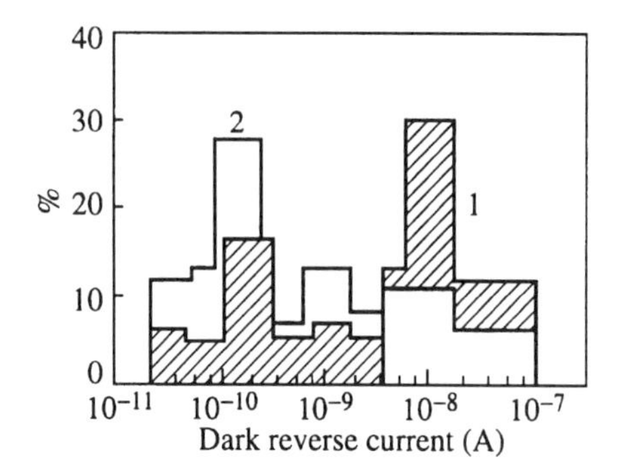

Figure 7.3. Histogram of the leakage current distribution for mesastructures from the same wafer: (1) Before the passivation; (2) after the passivation.

decreasing leakage current was caused by uncontrolled native oxides dissolving at the surface (among these P_2O_5 and As_2O_3) in the polyamide acid.

Accelerated thermotests were used on mesastructures passivated with the polyimide using minimal values for the currents (see Fig. 7.2, curve 1). The tests lasted 500 hours with $U = 10$ V and $T = +65°C$. No appreciable change in the current (at the level of 2×10^{-10} A) was observed.

Hence the results achieved using polyimide passivation are as good as those obtained by passivation with inorganic dielectrics, and in some cases, even better. This process can be used in industry.

The technology of the polyimide passivation developed by the Department of Semiconductor Heterostructure Physics of the Ioffe Physico-Technical Institute of the Russian Academy of Sciences has reached a level at which industrial applications become feasible.

7.2. PASSIVATION OF SURFACES THAT ARE ACTIVE ELEMENTS OF THE DEVICES (TYPE B DEVICES PASSIVATION)

The problem of passivation of the devices based on InP and InGaAsP with the passivated surfaces being an active element of the device has been investigated for more than 15 years. During this time it became possible to reduce considerably the density of the surface states N_{ss} on the passivated surfaces and to acquire a better understanding of the processes occurring on the surface.

The main problem that still remains unsolved is the problem of the time instability of the InP (InGaAsP)/insulator interface.

The achieved minimal values of N_{ss} are about $(2-5) \times 10^{11}$ $eV^{-1}cm^{-2}$ (Akazawa et al. 1989; Hollinger et al. 1990) and even $(5-7) \times 10^{10}$ $eV^{-1}cm^{-2}$ (Belyakova et al. 1992; Gardner et al. 1984).

Despite such small values of N_{ss}, the time stability of the surface parameters can be kept only during less than 10^3-10^4 s (Lile and Taylor 1983; Sawada et al. 1984; Van Vechten and Wager 1985; Juang et al. 1988; Akazawa et al. 1989; Tardy et al. 1991). At the same time the surfaces for the type A devices, passivated in the same way, provide a long and stable operation of the devices for thousands of hours.

The main causes of the III-V compounds-insulator interface instability in the type B devices were analyzed as far back as the 1980s. Then attention was paid to the processes going on in the near-surface defect layer, native oxide, and deposited insulator. These regions are shown in Figure 7.4.

Some authors (Lile and Taylor 1983; Hollinger et al. 1985) have suggested that the main role was played by the thermally activated electron tunneling into discrete traps in the uncontrolled layer of the native oxide between InP (InGaAsP) and the deposited dielectric. Also they have observed that the phase composition of the native oxides of InP (InGaAsP) is rather complex and contains thermodynamically unstable phases, even those similar to the semiconductive phases (Hollinger et al. 1985; Schwartz et al. 1982). Properties of the native oxide on InP were studied quite thoroughly (Hollinger et al. 1985; Ishill et al. 1989). It was shown that the

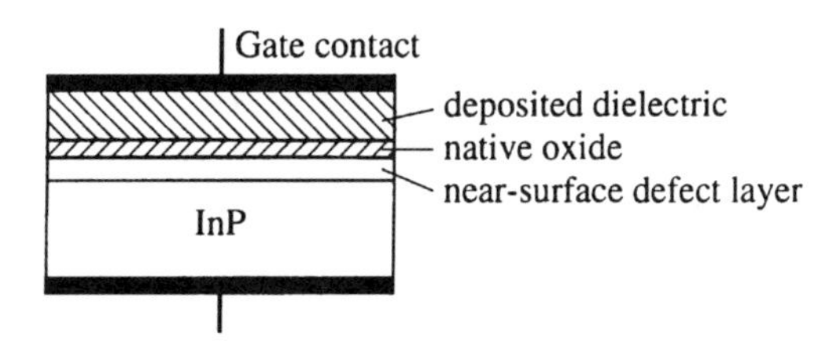

Figure 7.4. Cross-sectional view of a metal-dielectric-semiconductor structure.

polyphosphates $In(PO_3)_2$ are one of the most thermodynamically stable phases of the native oxide. New methods of anodic oxidation of InP have been developed that have made it possible to form the native oxide with an increased content of $In(PO_3)_2$.

The surface state density in such structures could thus be reduced, compared with the usual regimes of forming anodic oxides (up to 2×10^{11} eV^{-1} cm^{-2}). However, the time stability, estimated by the flat-band voltage shift, was low, lasting only for 4×10^2 s (Tardy et al. 1991).

Other authors (Van Vechten and Wager 1985; Juang et al. 1988; Hasegawa and Sawada 1983) have reasoned that the traps responsible for the instability were associated, not with the native oxides, but with the near-surface defects in InP (InGaAsP) that appeared during the dielectric formation. Van Vechten et al. (1985) suggested that the phosphorus vacancy may be such a defect and proposed a mechanism for trapping electrons by the shallow acceptors created by the movement of a nearest-neighbor In atom into the phosphorus vacancy.

A more general approach to understanding the time instability of the interface parameters was given in several publications (Hasegawa and Sawada 1983; Sawada et al. 1984; Akazawa et al. 1989; Hasegawa et al. 1989; Hashizume et al. 1994; Kodama et al. 1995). These authors suggested that the instability is caused by a high density of the interface states at the insulator-semiconductor interface. Interface states originate from bonding configuration disorder in the interface region due to random stress, loss of stoichiometry, and interface irregularity.

These publications stimulated studies on improving the methods of passivation in two directions:

1. Investigations of the properties of native oxides and of the parameters of the interface of InP (InGaAs) dielectric led to the development of methods for obtaining native oxides with controlled composition as well as two-layer dielectric structures containing the native oxide with a controlled composition.
2. Investigations on preventing excess phosphorus vacancies led to the development of processing semiconductor surfaces in a

phosphorus vapor or introducing a thin (5–10 Å) epitaxial layer of silicon in order to prevent interaction between the growing dielectric layer and InGaAs (InP).

At the Laboratory of Quantum-Size Heterostructures of the Ioffe Institute, a new method for forming a native oxide with controlled composition in oxygen plasma (Belyakova et al. 1992) was proposed. Unlike the anodic oxidation technique, this method could combine in the same process the formation of the native oxide and the application of a second dielectric (SiO_2, Si_3N_4, Ta_2O_3, Al_2O_3).

Our investigations showed that the plasma oxidation of InP causes changes in the structure and composition of the formed native oxide. This was proved by the changes observed in the refractive index n of the native oxide, depending on the exposure of InP in the oxygen plasma (Fig. 7.5, curve 1).

Three changes typically occur in the refractive index that correspond to the three stages of native oxide formation. During the first stage ($t_0 \leq 15$ min), an abrupt change of the growth velocity and of the refractive index is observed (Fig. 7.5, curve 2).

In the second stage ($15 < t_0 < 35$ min), the growth rate markedly slows, and the values of n become stabilized. Finally, in the third stage ($t_0 > 35$ min), the growth velocity is practically constant.

An investigation of the IR transmission spectra of the native oxide obtained at these three stages, within the range of 400–1500

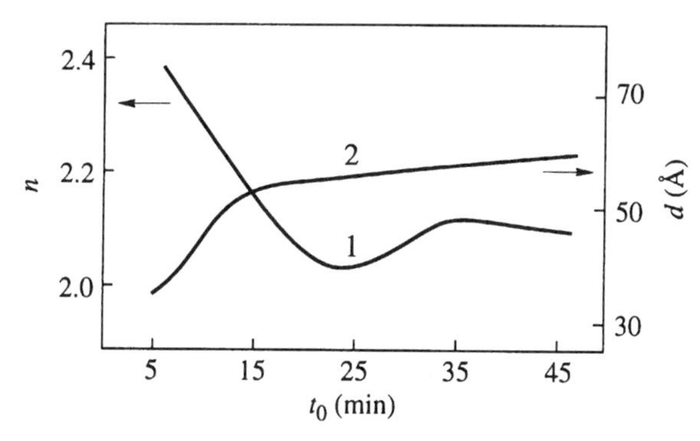

Figure 7.5. Dependence of refractive index n (curve 1) and thickness d (curve 2) of plasma native oxide on the time of plasma oxidation t_0.

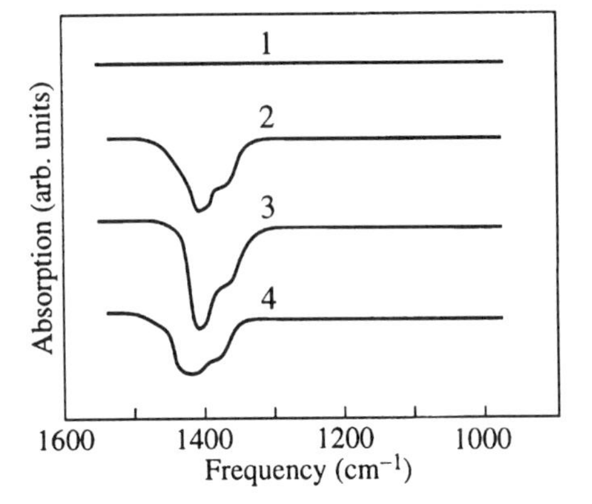

Figure 7.6. Infrared spectra for native oxides on InP: (1) Thermal or native oxide; (2–4) after the plasma oxidation during 25, 15, and 35 minutes, respectively.

cm^{-1}, shows a wide absorption band of plasma oxides observed in the region of 1400 cm^{-1} (Fig. 7.6, curves 2, 3, and 4). As a rule, this band is not observed for thermal oxides or for natural native oxides (Fig. 7.6, curve 1).

According to Pechkovskiy et al. (1991), this band may be linked to the vibration of the P=O groups in the $P_4O_{10}(PO_{3n-x})_x$-groups as well as to the stretching vibrations O — P — O in these groups. This indicates the presence of the polymer phase $In_xP_yO_z$. The maximum intensity of this band is observed for the native oxides obtained in the second stage of the process (curve 2, Fig. 7.6).

An investigation of *C-V* characteristics of the MIS structures based on the InP plasma oxides has shown that the minimal values of N_{ss}($\sim$ (4–5) $\times$ 10^{10} eV^{-1}cm^{-2}) and the smallest hysteresis loop of the *C-V* characteristics (0.2–0.3 V) are observed for native oxides, obtained after the second stage of oxidation (Fig. 7.7, curve 2). For comparison, the same figure shows a calculated *C-V* dependence (curve 1) and the *C-V* characteristic of the sample obtained after the third stage of the oxidation process (curve 3).

The samples obtained after different stages of the oxidation process and with a different content of the polyphosphate phase differ quite noticeably in the time stability of their interface parameters. For the specimens obtained after the first stage, the changes in the *C-V* characteristics are observed even during the measurement process. The most stable in time are the interface parameters after the second stage of the process.

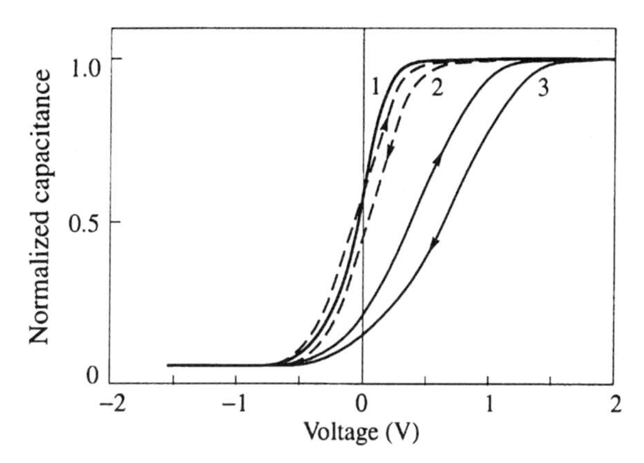

Figure 7.7. *C-V* curves for MIS structures on the basis of InP−plasma oxide (C_0 is the capacitance of dielectric): (1) Theory; (2) after 25-min oxidation; (3) after 35-min oxidation.

The stability estimations obtained from measuring the shift of flat band for the MIS indicate the stability of interface parameters up to time $\tau_s \sim 10^3$ s. These values (as well as the obtained N_{ss} values) are more stable than those for anodic oxides with a controlled composition (Tardy et al. 1991). However, for the reliable functioning of the devices, these values are still not sufficient. A better stability ($\tau_s \sim 10^4$ s) has been observed in the paper (Hollinger et al. 1990), where the process of annealing under As vapor pressure, followed by a controlled oxidation, was used for the passivation.

Record values of $\tau_s \sim 5 \times 10^4$ s were obtained by using thin silicon layers (5−10 Å) grown on an InGaAs surface using the MBE process, accompanied by oxidation via a photochemical vapor deposition (CVD) technique (Akazawa et al. 1989).

The results given here show that despite all efforts no qualitative leap was made in solving the problem of the time stability of the dielectric III-V semiconductor interface. No revolutionary approach to the solution of this problem has emerged, even though the concept concerning the nature of surface states developed in some papers was quite promising (Hasegawa et al. 1989; Hashizume et al.

1994; Kodama et al. 1995). Therefore it seems reasonable to suggest other directions for further investigation.

Si technology has shown that a qualitative leap in improving insulator-Si interface parameters can be achieved when the processes of generation and relaxation of the nonequilibrium intrinsic defects are taken into consideration (Ivanov et al. 1982; Sze 1983).

For III-V compounds, in particular, for InP and InGaAsP, the initial stage of disordering of the near-surface regions and, especially, the dynamics of this process have been poorly understood. It is possible that the study of the transfer conditions from the nonequilibrium state into the equilibrium state will provide the key to solving the instability problem. First of all, one has to understand thoroughly the process of generation and relaxation of nonequilibrium intrinsic defects taking place when the interface is formed.

All the methods of forming an insulator on InP (InGaAsP) are low-temperature methods. Under such conditions it is difficult to provide an effective relaxation of nonequilibrium defects, not only in the near-surface region of a dielectric but also in the bulk.

Even for Si it has been impossible to obtain a low-temperature SiO_2 with the same small density of states of the interface and the same good time stability as for the thermal SiO_2 (Inoue et al. 1989; Bashkin et al. 1990). As a rule, for the Si–low-temperature SiO_2 interface, the values of $N_{ss} \sim 10^{11}$ $eV^{-1}cm^{-2}$ are typical, along with the large hysteresis of the *C-V* characteristics and the negative fixed charge. As a rule, the data on the time stability are not quantified. Our investigations of the Si–low-temperature SiO_2 interface have shown that the time stability of the parameters for such an interface does not exceed $\tau_s \sim 10^3$ s. The typical *C-V* characteristics of these structures are given in Figure 7.8, curve 1 (compare it with curve 2 of Fig. 7.7).

Also the final cleaning of the wafers used in the technology of InP and of InP related compounds is not optimal. The experimental proof is illustrated by a comparison of curves 1 and 2 in Figure 7.8. In this experiment a layer of SiO_2 was grown on two Si wafers by the same CVD method. One of the wafers was cleaned according to the standard Si technology before spinning the oxide. The *C-V* characteristic for that wafer is given by curve 1. The other wafer was cleaned in addition with the agents used in the InP technology

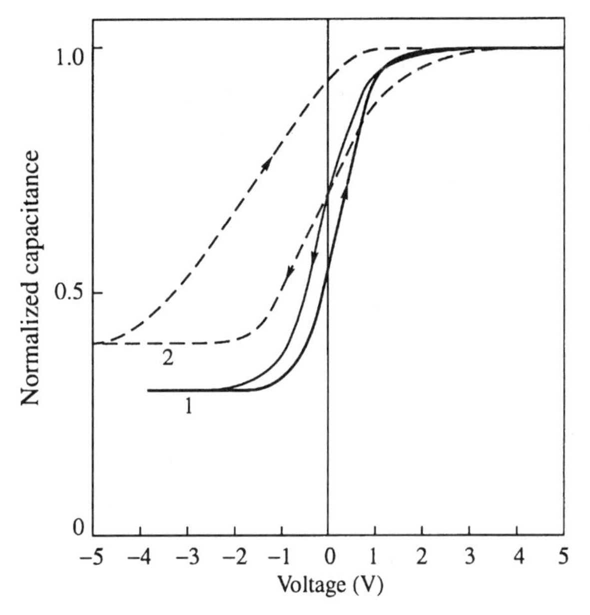

Figure 7.8. Typical *C-V* curves of MIS CVD Si-SiO_2 structures: (1) Standard Si cleaning; (2) standard Si cleaning plus the cleaning in chemical agents used in the InP technology.

(curve 2). As can be seen, in the latter case the loop width of the hysteresis increases quite notably. This shows that apart from such peculiarities of III-V compounds as a strong disordering of the near-surface region of the semiconductor in the process of dielectric forming and a possible presence of an uncontrolled composition on the interface, the contribution to the time instability can be also caused by the nonequilibrium methods of obtaining the insulators and by the nonoptimal preparation of the InP and InGaAsP surfaces.

7.3. CONCLUSION

We have demonstrated an effective use of polyimides for the passivation of the surfaces that are not active elements of InP (InGaAsP) devices (e.g., p-n junction peripheries). The main problem that remains unsolved for the devices where the passivated surface is an active element is the time instability of the parameters of insulator-InP (InGaAsP) interface. Many necessary conditions of the passivation of such surfaces have become clear, in particular, the formation of the intrinsic oxide of the controlled composition and the prevention or treatment of the semiconductor near-surface region disordering beneath the dielectric. The modern level of technological development allows one to achieve these conditions. Still the time stability of the interface parameters remains unacceptable and prevents the practical use of the passivated surfaces for microelectronic devices.

There remains important work to be done in analyzing more thoroughly the behavior of nonequilibrium defects in the near-surface region of the semiconductor and insulator grown by nonequilibrium low-temperature methods. Studying the influence of such defects on the parameters of the InP (InGaAsP)-insulator interface and on their time stability will facilitate the solution of the time instability problem.

REFERENCES

Akazawa, M., Hasegawa, H., and Ohue, E. (1989). *Jap. J. Appl. Phys.* 2, *Lett.* 28, L2095–L2097.

Bashkin, M. O., Emel'yanov, A. V., Men'shikov, O. D., Portnov, S. M., and Ufimtsev, V. B. (1990). *Electron. Tekh.* 3, *Mikroelektronika* 19, 32–35 (in Russian).

Bauer, G. J. and Trommer, R. (1988). *IEEE Trans. Electr. Devices* ED-35, 2349–2353.

Belousov, M. V., Gorelenok, A. T., Davydov, V. Yu., et al. (1990). *Sov. Phys. Semicond.* 24, 1349–1351.

Belyakova, E. D., Ber. B. Ya., Gorelenok, A. T., et al. (1989). *Proc. 1st All-Union Conf. on Physical Basics of Solid-State Electronics*, Leningrad, Russia, pp. 238–239 (in Russian).

Belyakova, E. D., Gabaraeva, A. D., Gorelenok, A. T., et al. (1992). *Poverkh., Fiz. Khim. Mekh.* 7, 88–93.

Biernath, R. W. and Soane, D. S. (1989). *PME '89, Polymers for Microelectronics Science and Technology*, pp. 104–105. Tokyo, Japan.

Chané, J. P., Martin, B. G., Patillon, J. N., and Gentner, J. L. (1986). *Le vide les couches minces*. 41, 203–205 (in French).

Gardner, P. D., Narayan, S. Y., and Yun, Y.-H. (1984). *Thin Sol. Films* 117, 173–190.

Goncharova, T. S. (1989). *Zarubezhnaya Electronnaya Tekhnika* 8, 53–82 (in Russian).

Gorelenok, A. T., Mamutin, V. V., Sulima, O. V., et al. (1985a). *Sov. Phys. Semicond.* 19, 668–673.

Gorelenok, A. T., Il'inskaya, N. D., Korol'kov, V. I., Mamutin, V. V., and Shmidt, N. M. (1985b). *Sov. Phys. Tech. Phys.* 55, 1568–1571.

Hasegawa, H. and Sawada, T. (1983). *Thin Sol. Films* 103, 119–140.

Hasegawa, H., Ohno, H., Ishii, H., et al. (1989). *Appl. Surf. Sci.* 41–42, 372–382.

Hashizume, T., Hasegawa, H., Riemenschneider, R., and Hartnagel, H. L. (1994). *Jap. J. Appl. Phys.* 1, *Regul. Pap. Short Notes*. 33, 727–733.

Hollinger, G., Bergignat, E., Joseph, J., and Robach, Y. (1985). *J. Vac. Sci. Technol. A*, *Vac. Surf. Films* 3, 2082–2088.

Hollinger, G., Blanchet, R., Gendry, H., et al. (1990). *J. Appl. Phys.* 67, 4173–4182.

Inoue, K., Okuyama, M., and Hamakawa, Y. (1989). *Appl. Surf. Sci.* 41–42, 407–410.

Ishii, H., Hasegawa, H., Ishii, A., and Ohno, H. (1989). *Appl. Surf. Sci.* 41–42, 390–391.

Ivanov, E. T., Sukhanov, V. L., Tuchkevich, V. V., and Shmidt, N. M. (1982). *Sov. Phys. Semicond.* 16, 129–131.

Johnson, G. A. and Kapoor, V. J. (1991). *J. Appl. Phys.* 69, 3616–3622.

Juang, M. T., Wager, J. F., and Van Vechten, J. A. (1988). *J. Electrochem. Soc.* 135, 2019–2023.

Kodama, S., Koyanagi, S., Hashizume, T., and Hasegawa, H. (1995). *Jap. J. Appl. Phys.* 1, *Regul. Pap. Short Notes* 34, 1143–1148.

Le Bellego, Y., Blanconnier, P., and Praseuth, J. P. (1990). *Rev. Phys. Appl.* 25, 941–945 (in French).

Lile, D. L. and Taylor, M. J. (1983). *J. Appl. Phys.* 54, 260–267.

Pechkovskiy, V. V., Mel'nikov, R. Ya., and Dzyuba, E. D. (1981). *Atlas Infrakrasnykh Spectrov Fosfatov, Ortofosfaty*, Nauka, Moscow, 247 (in Russian).

Sawada, T., Itagaki, S., Hasegawa, H., and Ohno, H. (1984). *IEEE Trans. Electron. Devices* ED-31, 1038–1043.

Schwartz, G. P., Sunder, W. A., and Griffiths, J. E. (1982). *J. Electrochem. Soc.* 129, 1361–1367.

Soone, D. S. and Martynenko, Z. (1989). *Polymers in Microelectronics: Fundamentals and Applications*. Elsevier, Amsterdam.

Sze, S. M. ed. (1983). *VLSI Technology*., McGraw-Hill Book Company, Tokyo, Toronto.

Tardy, J., Thomas, I., Viktorovich, P., et al. (1991). *Appl. Surf. Sci.* 50, 383–389.

Van Vechten, J. A. and Wager, J. F. (1985). *J. Appl. Phys.* 57, 1956–1960.

Yeats, R. and Von Dessonneck, K. (1984). *Appl. Phys. Lett.* 44, 145–147.

CHAPTER 8

PRECISION PROFILING OF SEMICONDUCTOR SURFACES BY PHOTOCHEMICAL ETCHING

D. N. GORYACHEV, L. V. BELYAKOV, and O. M. SRESELI

Some of the first semiconductor lasers with small divergence of radiation (Alferov et al. 1975) and distributed feedback (Alferov et al. 1976b) were fabricated at the Ioffe Institute in St. Petersburg by using the maskless precision photochemical etching technique developed for semiconductors by the authors of this chapter.

Precision profiling of semiconductor surfaces developed as a result of the rapid growth of holography and optoelectronics. Holograms, reflective holographic diffraction gratings for spectroscopic devices, and relief diffraction gratings, as both passive and active elements of optoelectronics, all require the profiling of surfaces of the solids, and this list is far from complete.

In solving those tasks, two processes have been generally applied: the photolithographic process, widely used in the modern microelectronic industry, and the method of maskless photoetching of semiconductors, which is much less developed. Each technique has certain merits not found in the other.

Maskless photoetching of semiconductors is based on affecting their surface by illumination, which is nonuniform over the area and by applying a certain etchant simultaneously. Two relatively

Semiconductor Technology: Processing and Novel Fabrication Techniques,
Edited by M. Levinshtein and M. Shur.
ISBN 0-471-12792-2

independent directions of investigation can be specified that have a common theoretical basis: *photoelectrochemical* (mostly anodic) *etching* and *photochemical etching*, which does not require the application of the external electric field. The latter, which is the most practical, is considered in this chapter.

In the earliest publications on photochemical etching of semiconductor surfaces (Si, PbS), a transparency was projected and a relief metal-coated image of the projected object appeared on the semiconductor surface (Goryachev et al. 1970a, 1970b). At the same time it was established how to obtain a precision relief on the silicon surface by photoelectrochemical method (Dalisa and DeBitetto 1970a; Dalisa et al. 1970b). The same publications drew attention to the fact that on a silicon surface the relief can also be formed in a mixture of certain acids without the application of an external bias. But in that case the results were much worse. Later the photoelectrochemical method was studied thoroughly and was perfected at the Institute of Semiconductors of the Ukrainian Academy of Science by a research group headed by V. A. Tyagai. They also discovered certain limitations to the method: It was impossible to profile successfully p-type semiconductors and high-resistance semiconductors. Also there was an incomplete dissolution of all the components of semiconductors with complex chemical composition, and so on. As shown below, the photochemical etching is, to a great extent, free of such drawbacks.

A thorough development of the theory of photochemical etching as well as the development of certain technological processes for a great number of semiconductors were performed at the Ioffe Institute. The results of these investigations were published in 1974 and later by the authors of this chapter (Belyakov et al. 1974a, 1974b), and they were partially included into review papers (Gurevich and Pleskov 1983; Rytz-Froidevaux and Salathe 1985).

8.1. PHYSICS AND CHEMISTRY OF PHOTOCHEMICAL ETCHING OF SEMICONDUCTORS

8.1.1. Mechanisms of Photochemical Etching

The chemical etching of semiconductors, as well as of many other solids, requires a certain energy to break the interatomic bonds in the crystal lattice. In the semiconductor-electrolyte system two relevant sources of that energy are the energy of light, incident on the semiconductor when it is illuminated, and the energy of a chemical reaction between the semiconductor and the solution.

As an example of etching whose main source of energy is the energy of light, we can name processes governed by the *photovoltaic* mechanism. Their driving force is the voltage that appears between the illuminated and unilluminated parts of the surface of the semiconductor. It is well known that an interfacial potential exists at the semiconductor-solution interface. The electric field penetrates into the semiconductor and forms a space-charge region (Myamlin and Pleskov 1967). When the semiconductor is illuminated, the photogenerated electron-hole pairs are separated by the space-charge field, and a photovoltage appears between the surface and the bulk of the semiconductor. The sign of the photovoltage is determined by the type of the conductivity of the semiconductor. The magnitude (within certain limits) is proportional to the illumination intensity. The photovoltage is zero at the unilluminated parts of the semiconductor surface. As a result a voltage drop is established between the illuminated and unilluminated (or weakly illuminated) parts of the surface. Since the whole of the semiconductor surface is in contact with the conducting solution, circular electric currents appear that flow through the solution and the bulk of the semiconductor, as is shown in Figure 8.1. As this process takes place, parts of the semiconductor surface with a greater positive potential can, under certain conditions, dissolve in accordance with the electrochemical (anodic) mechanism (Haisty 1961; Kuhn-Kuhnenfeld 1972). Simultaneously the parts with a less positive (or a more negative) potential experience a cathodic reduction of the solution components, for example, hydrogen ions.

In photoetching processes governed by the *corrosion* mechanism, another dissolution mechanism is observed (Belyakov et al. 1976b). In solutions that are oxidants (i.e., acceptors of electrons) with respect to the given semiconductor, etching takes place owing to the chemical energy released during the oxidation of the semiconductor. The oxidant permanently takes electrons from the semiconductor, which results in "loosening" the interatomic bonds within the semiconductor and eventually leads to its dissolution. When being dissolved, the semiconductor supplies the solution with positively charged ions, losing a certain number of holes with every ion.

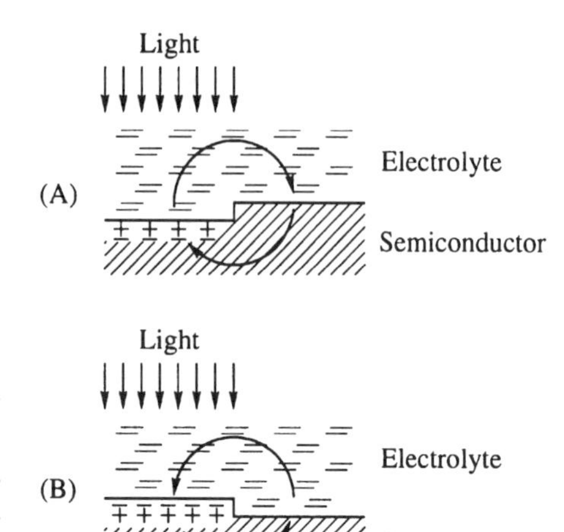

Figure 8.1. Direction of circular currents for photovoltaic mechanism of etching: (A) n-type semiconductor; (B) p-type semiconductor.

Thus equal quantities of electrons and holes are transferred to the solution simultaneously. As a result the electrical neutrality of both the solution and semiconductor is maintained, and the dissolution of the semiconductor goes on quite steadily. The dissolution rate is determined by the thermal generation rate of the minority carriers and is usually not high. It is the so-called *dark rate* of etching. With the illumination of the semiconductor by a strongly absorbed light, both the concentration of minority carriers and the rate of etching increase, irrespective of semiconductor conduction type. The light energy in this corrosion process leads to an additional weakening of the interatomic bonds.

In a number of cases, the mechanism of the semiconductor etching may coincide with that of the corrosion, even though the solutions do not contain any oxidant. Processes of the photochemical etching of Cu_2O and CdS in hydrochloric acid solutions (Goryachev and Paritskii 1974) are typical in this respect. The above-mentioned semiconductors are compounds with a considerable part of ionic component in interatomic bonds. The contact of such compounds with polar solvents leads to the weakening of ionic bonds and to the dissolution of the semiconductor. Illumination weakens the interatomic bond even more and accelerates the dissolution. As this takes place, the atoms of semiconductor get fully ionized and are transferred to the solution as solvated ions (Belyakov et al. 1979c). Since the solvation energy of ions is the driving force of such processes, the whole process gets the name *solvation* (or nonoxidizing) photochemical etching. Naturally such a process of dissolution is impossible if a semiconductor possesses purely covalent interatomic bonds (silicon, germanium).

Now we will dwell on some important distinctions among the above processes.

Photovoltaic etching has the following characteristics:

1. The etching of a semiconductor takes place provided that the illumination of its surface is nonuniform. Otherwise, the rate of etching is equal to zero if the illumination is uniform or if there is no illumination at all.
2. Since the direction of the circular currents is determined by the type of the conductivity of the semiconductor, in the *n*-type semiconductors the illuminated parts of the surface are dissolved, and in the semiconductors of *p*-type the darkened parts get dissolved.
3. Only the anodic, that is, positively charged, parts of the semiconductor surface are subjected to etching, while the cathodic parts do not change at all. So the formed relief can be quite different from the optical picture being projected upon the semiconductor. For instance, with a sinusoidal light distribution, typical for holographic images, the formed relief has periodically located concave grooves.
4. The density of circular currents and, consequently, the rate of the photovoltaic process are inversely proportional to the resistivity of the semiconductor and that of the solution. Therefore, in practice, the acceptable rate of etching can be achieved only for comparatively low-resistance semiconductors (Goryachev et al. 1975).

Corrosion and *solvation* mechanisms are characterized as follows:

1. There is always a certain *dark* rate of dissolution that is determined by the equilibrium concentration of minority carriers as well as by some other processes, such as the injection of holes by the oxidant.

2. Since the concentration of minority carriers is, within certain limits, proportional to illumination, the shape of the relief being formed reproduces the light distribution on the semiconductor surface.
3. The etching rate is always higher for the illuminated parts of the semiconductor, irrespective of its conductivity type.
4. The etching rate does not depend on the resistivity of the semiconductor or solution because both electrons and holes escape into the solution on the same parts of the surface. This enables successful profiling of the high-resistance semiconductors whose resistivity reaches 10^{10}–10^{12} $\Omega \cdot$ cm.

The relative contribution of each mechanism can be quite different, depending on the physicochemical properties of both the semiconductor and the etchant. Photochemical etching is primarily used in order to fabricate reflective gratings. They are etched on the semiconductor surface when illuminated by two coherent laser beams forming an interference field with a sinusoidal light distribution. For the sinusoidal relief the best etching should follow the corrosion or solvation mechanism.

8.1.2. Composition of Etching Solutions

The problem of the chemical composition of the etching solution is directly linked to realization of etching mechanisms. As was mentioned before, in recording holograms and fabricating diffraction gratings, it is preferable to apply solutions that provide a corrosion or solvation etching mechanism. Below we will formulate additional requirements for such etching solutions:

1. To achieve high photosensitivity of the process, the kinetics of etching is to be determined by the rate of carrier generation by light in the space-charge region, and not by the rate of diffusion in the solution. Therefore it is desirable to apply sufficiently concentrated solutions that have low viscosity.

2. Oxidants are known to consume electrons from the semiconductor conduction band, and in some cases there is even a simultaneous capture of electrons from the valence band. The latter effect

is largely linked to the chemical nature of the oxidant (Pleskov 1960). The effect corresponds to the injection of holes into the semiconductor, which enhances the dark dissolution rate and reduces the light sensitivity of the process. That condition reduces the number of oxidants.

3. For high resolution etching, it is necessary to avoid the evolution of gases, which will distort the optical image, since the semiconductor is illuminated through the solution. This precludes the application of such "classical" oxidants as nitric acid or hydrogen peroxide.

4. The electrolyte must be transparent for the wavelength of the light source. At the same time, to create a light interference field with a small period, the light sources must correspond to the short-wavelength region of the spectrum. Hence a large group of oxidants whose solutions are yellow, orange, or red in color cannot be used. This group includes such oxidants as chromates and bichromates, permanganate, iodine, and bromine among many others.

5. The products of a chemical reaction must not form insoluble compounds. That can occur in introducing into solutions certain complexing agents, acidity regulators, and the like, that make the composition of the solutions even more complicated.

6. All other factors being equal, it is necessary to use etchants with high oxidation activity capable of dissolving the illuminated semiconductor at a significant rate. Thus choosing the proper chemical composition of the etching solution is a complicated problem, demanding a lot of experimental investigations for each semiconductor. The difficulty is later compensated by the relative simplicity of the technological process after it has been developed.

8.1.3. Resolution of the Process and the Demands on Semiconductors

Even the earliest publications on photochemical etching noted the discrepancy between the high resolution of the process and the long diffusion length of carriers in semiconductor materials. One might expect that diffusion should have led to a considerable spreading of carriers in the surface region of the semiconductor. The evaluation of the electric field in the space-charge region has shown that contrary to the results of Dalisa and DeBitetto (1970a), the fields value is not sufficient for the successful extraction of minority carriers at the surface of a semiconductor, and it cannot explain such a small spreading (Belyakov et al. 1978). A mechanism explaining that discrepancy and showing a way of improving the method was proposed by Belyakov et al. (1976a). We will limit ourselves to a qualitative consideration of this problem for the corrosion mechanism of etching.

The effective spread length of the carriers, created by light in a certain region of a semiconductor, is defined by the rate of their recombination in the bulk and on the surface. By enhancing the recombination intensity, it is possible to reduce this length. But then the surface concentration of minority carriers inevitably decreases, as consequently does the photoetching intensity.

The process of etching itself acts as an additional mechanism, leading to the removal of carriers. As was already mentioned, in the process of etching, equal numbers of electrons and holes leave the semiconductor for the solution. That phenomenon suggests an additional channel of carrier recombination at the interface of the semiconductor and the etchant. This kind of recombination, which has no analogies in solid-state electronics, is called *electrochemical recombination.* It is evident that with the increase of etching rate, the intensity of such a recombination increases, and the process resolution improves as well. Calculations that have considered all recombination channels for low-resistance gallium arsenide have shown that a relief period can be reduced down to 30 nm (Belyakov et al. 1979a).

We can now formulate some general physicochemical facts that can lead to optimal results in photochemical etching.

1. Highly doped, but not degenerate, semiconductors will minimize the concentration of the equilibrium minority carriers which determine the dark etching rate, and result in a higher-quality semiconductor etched surface.

2. The energy band bending of n-type semiconductors, caused by contact with oxidants, is in the same direction as "natural" blocking

band bending. So the corresponding electric fields in the space-charge region add up, which improves the process resolution.

3. A strongly absorbed light or, for a given wavelength of light, a semiconductor with a maximum light absorption coefficient will diminish the minority carrier spread and thus improve the resolution of the process.

8.1.4. Correlation of Relief Shape and Light Distribution

The corrosion mechanism indicates that the surface concentration of minority carriers and, consequently, the semiconductor etching rate must be proportional to illumination intensity in a wide range of intensities. However, deep diffraction gratings often deviate from the purely sinusoidal shape of the relief.

A computer simulation of the etching process in the interference field was performed at the Ioffe Institute. The etching rate at a given point of the surface was assumed to be proportional to the electromagnetic field intensity at that point, and etching was assumed to be normal to the surface. Our calculations for gratings with periods of 0.25, 0.8, and 2.5 μm have enabled us to draw the following conclusions:

- Distortion of the relief's shape depends on the relative values of the relief's depth, namely on the depth-to-period ratio of the grating ($\mathbf{h}/\mathbf{a}$). The distortion only weakly depends on the relief's period.
- Total amplitude of the higher harmonics of the relief does not exceed 1 percent of the amplitude of the main harmonic for $\mathbf{h}/\mathbf{a} < 0.05$ and 10 percent for $\mathbf{h}/\mathbf{a} < 0.1$.
- Retardation of the first relief's growth harmonic takes place beginning with $\mathbf{h}/\mathbf{a} > 0.2$, and with $\mathbf{h}/\mathbf{a} > 0.3$ the growth of the first harmonic practically stops. This effect was observed experimentally (Belyakov et al. 1979b).

Figure 8.2 illustrates the change in relief's shape and depth over time for a sinusoidal grating with a period of 0.8 μm. It can be seen that there are appreciable relief distortions from the sinusoidal

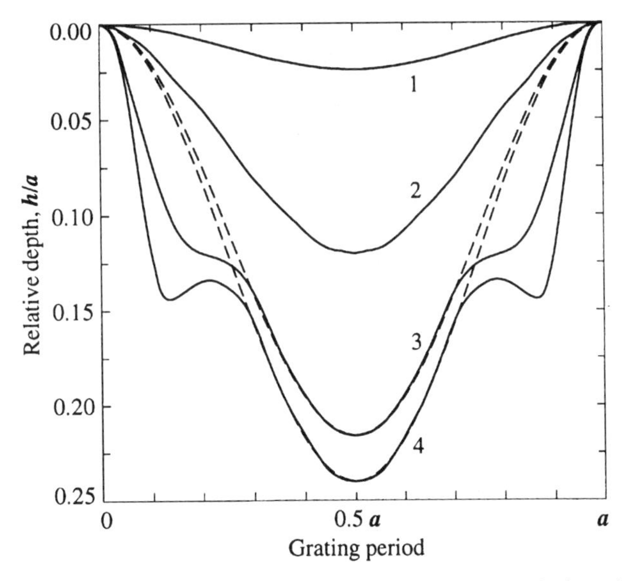

Figure 8.2. Changes of the relief shape during long-period etching (computer simulation). Etching times (min): (1) 2; (2) 20; (3) 36; (4) 40. The dashed lines give the sine-shaped profile.

shape for $\mathbf{h/a} > (0.20-0.25)$. At the largest achievable depth of the grating, $\mathbf{h/a} = (0.3-0.5)$, the distortion is quite noticeable.

The discussion above refers only to isotropic etching of semiconductors, such as vitreous chalcogenides (Belyakov et al. 1977) or polycrystalline films (Goryachev et al. 1970b). Monocrystalline semiconductors have very different etching rates along different crystallographic axes. As a result there may be large distortions in the relief's shape, especially when the relief period is small (Matz and Zirrgiebe 1988). For instance, with a period of less than 1 μm, the grooves of the diffraction grating must be parallel to the direction of one of the main crystallographic axes of semiconductor, for otherwise the grating grooves will be "torn."

Anisotropic etching is quite useful in a number of cases, however, because it enables one to fabricate special asymmetric gratings, concentrating light mostly in one direction (Egorov et al. 1984). Analogous results were obtained when, in the process of etching, the sample had a certain orientation relative to the incident laser beam (Podlesnik et al. 1983).

8.2. PRACTICAL PHOTOCHEMICAL ETCHING

Photochemical etching is used primarily for manufacturing reflective diffraction gratings with a period of order of the light wavelength in the visible part of the spectrum, which is less than 1 μm. They are etched on a semiconductor surface when it is illuminated by two coherent laser beams forming an interference field pattern with a sinusoidal distribution of illumination. Figure 8.3 gives a schematic view of the light geometry and the illumination distribution. If two beams of equal intensity fall upon the sample surface symmetrically, at an angle α to the normal to the surface in the absence of an etchant, the light intensity changes in accordance with the law

$$I(x) = I\left[1 + \cos\left(\frac{2\pi x}{\mathbf{a}}\right)\right], \tag{8.1}$$

where **a** is the interference field period, defined by the relation $\mathbf{a} = \lambda/2\sin\alpha$, where λ is the wavelength of light in the air.

8.2.1. Equipment Requirements

Holography diffraction gratings are known to be the simplest holograms. Standard holographic equipment is typically used for the photochemical etching experiment. The main elements are (see Fig. 8.4) a single-mode CW laser (1); a beam-splitter (2), for example, in the form of a semitransparent plate; telescopic beam-expanders (3); and several metal mirrors (4), for viewing the intersection of laser beams at the semiconductor's surface (5). The semiconductor is placed in a cell with the etchant (6). To control the quality of the

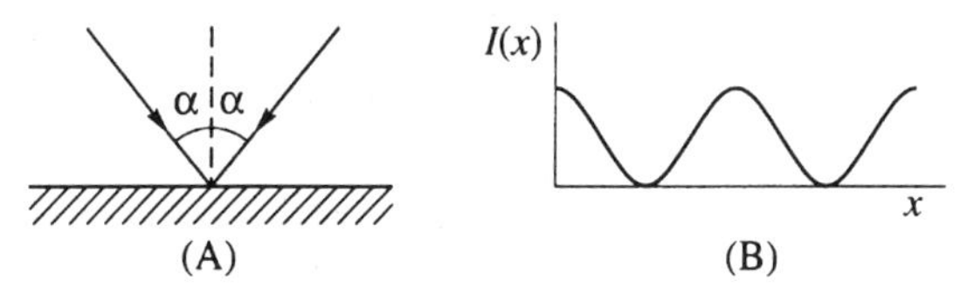

Figure 8.3. Schematics of the forming interference light field: (*A*) Geometry of light-beam incidence; (*B*) distribution of light intensity on the semiconductor surface.

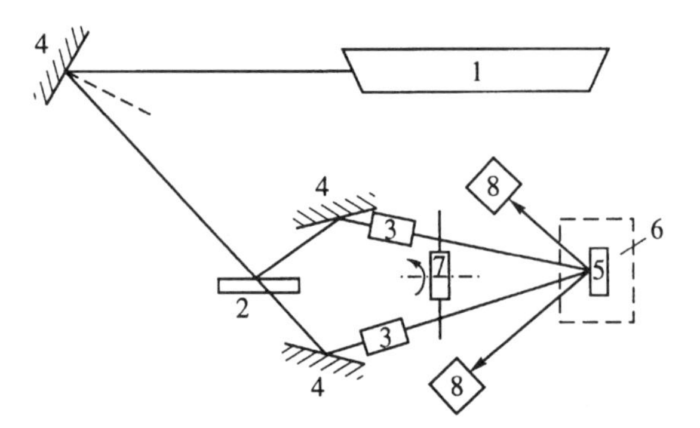

Figure 8.4. Holographic setup for the photochemical etching of diffraction gratings: (1) Laser; (2) beam-splitter; (3) telescopic beam-expander; (4) mirror; (5) sample; (6) cell with an etchant; (7) beam-chopper; (8) photodiode.

resulting grating, a mechanical beam-chopper (7) and photodetectors (8) are used; the latter record the intensity of light diffracting on the formed grating. The setup is mounted on a massive steel plate that rests on shock-absorbers.

There are many different types of cells used for photochemical etching. Some of them are shown in Figure 8.5. They are made of glass, quartz, or plastic so that they are transparent for the laser beam. Figure 8.5*A* shows the most commonly used rectangular cell positioned vertically. Another kind of cell is horizontal (Fig. 8.5*B*). Since it does not have a front wall, there is no interference of

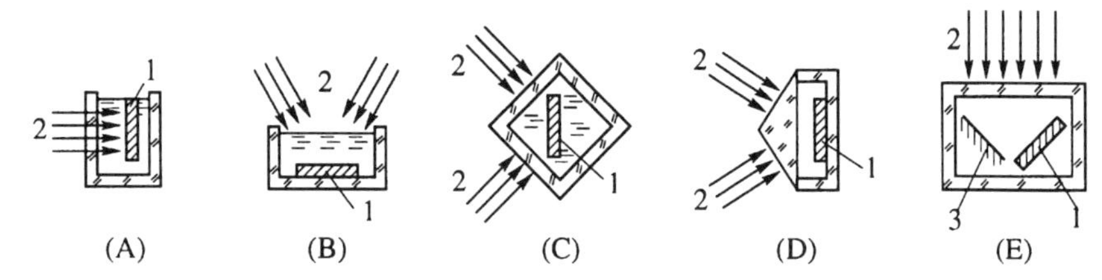

Figure 8.5. Cells for photochemical etching: Semiconductor (1); light beams (2). (*A*) Vertical construction, side view; (*B*) horizontal construction, side view; (*C*) vertical construction, with beams of light normal to the cell's windows, top view; (*D*) cell with an immersion prism, top view; (*E*) cell with a Lloyd mirror, top view.

reflected light. For both cells the grating period is defined by the expression (8.1). Hence the grating period cannot be less than $\lambda/2$ (with beams of light coming from the opposite directions). In reality, with the largest possible incidence angles, the grating period is larger than 0.6λ; that is, for $\lambda = 0.44$ μm the grating period cannot be less than 0.26 μm.

The cell shown in Figure 8.5*C* has entrance windows located perpendicularly to the light beams, so the entering light is not refracted. In this case the grating period is defined by

$$\mathbf{a}_1 = \frac{\mathbf{a}}{\mathbf{n}}, \tag{8.2}$$

where **n** is the refraction index of the etchant. The value of **n** can vary from 1.33 (for water) to 1.4 and more (for sulfuric acid solutions). The grating period decreases accordingly. Another version of such a cell is a cell that has an entrance immersion prism (Fig. 8.5*D*) whose refraction index coincides with that of the solution (Alferov et al. 1976a). In this case, by using the He-Cd laser with $\lambda = 0.44$ μm, it is possible to fabricate a grating whose minimal period is 0.16–0.17 μm.

Something special is the construction of another cell (Fig. 8.5*E*) with a Lloyd mirror located directly in the etching solution. Despite certain difficulties related to the choice of a corrosion-resisting material for such a mirror, this construction has better optical stability and additionally gives a "purer" image, since the cell contains a fewer number of optical elements located in the light.

A special technique that combines both photoresistive and photochemical etching allows one to double the inverse grating period. Such a technique was used to fabricate a diffraction grating with a period of 98.4 nm (Flanders et al. 1979).

There is also a "single-beam" method of forming a periodic relief. In a photochemical etching process the surface of the sample is illuminated by only one laser beam of a low intensity, incident perpendicularly to the surface. The grating period coincides with the wavelength in the solution -0.4 μm (Tsukada et al. 1983). That phenomenon is explained by the interference of the incident light with the waveguide mode adjacent to the sample layer of the etchant. The layer is supposed to have a refraction index different from its volume refraction index (Sychugov and Tulajkova 1984). This method has been used to produce diffraction gratings with relief periods from 0.3 to 6 μm and depths more than 1 μm (Golubenko et al. 1985).

8.2.2. Control of the Etching Process

Photochemical etching has a unique advantage: it allows one to observe and, if necessary, to regulate the formation of diffraction grating during its fabrication. In this technique the relief that forms on the semiconductor surface is simultaneously displayed, without the need for any further procedures. The diffraction of both beams, forming the interference field, takes place on the relief during the etching. When the interfering beams make small angles, namely when the grating period considerably exceeds the wavelength of light, it is possible to observe several diffracted beams located in the mirror reflection plane, symmetrically, on both sides of the normal to the semiconductor surface. Each of these beams is twice "degenerated"; that is, the diffraction maximum of the mth order from beam 1 is in the same direction as the diffraction maximum of the $m + 1$ order from beam 2 ($m = 0, \pm 1, \pm 2, \ldots$). In a short grating period, only one or two diffraction maxima are observed, but this is quite enough for effective etching control.

For small angles of incidence and diffraction, the relative intensity of light, diffracted by the reflective sinusoidal relief grating into the maximum of the mth order (i.e., the diffraction efficiency η_m), is proportional the square of the Bessel function of the first kind $J_m^2(\Gamma)$, where argument $\Gamma = 2\pi \mathbf{n}\mathbf{h}/\lambda$. Here $\mathbf{n}$ is the refraction index of the medium in contact with the grating, $\mathbf{h}$ is the total depth of the grating, and λ is the wavelength of light. In other words, the intensity of the diffracted light is determined by the depth of grating and by the order of the diffraction maximum. Assuming that the etching rate at a given light intensity is constant (which, within certain limits, has been confirmed experimentally), we can consider the dependence of diffraction efficiency on depth as a function of time (see Fig. 8.6, curve 1).

The method of controlling the etching process (Belyakov et al. 1979b) consists in a periodic "lifting degeneracy" by mechanically

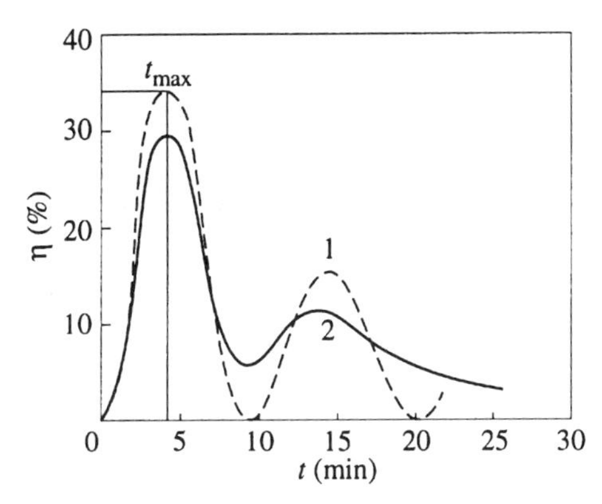

Figure 8.6. Changes in diffraction efficiency η in the process of interference etching: (1) Theoretic curve; (2) experimental curve (n-GaAs).

interrupting beams 1 and 2 for two to three seconds with a frequency of once or twice a minute. Using a photodetector monitoring the diffracted beam, the recorder fixes three values: the intensity of diffraction of the mth order, the intensity of the $m + 1$ order, and their total interference sum. An example of such a record is shown in Figure 8.7. Since the diffraction efficiency of the adjacent diffraction maxima changes differently with grating depth, the ratio of the diffraction efficiencies of these maxima is a function of depth. This makes it possible to work without measuring the absolute values of intensities of the diffracted light. A typical experimental dependence of diffraction efficiency on etching time is shown in Figure 8.6, curve 2. Recall that the surface of semiconductor is illuminated through the etchant with a refraction index **n**, and consequently the wavelength of diffracted light on the grating is **n** times lower than in the air. Therefore the diffraction efficiency observed during the etching does not correspond to the actual rate of formation of the relief but seems to be ahead of it. This makes it possible to estimate the depth of the obtained sinusoidal relief at any time with high accuracy. For instance, in order to fabricate a diffraction grating with a maximum diffraction in the first order, it is necessary after reaching the maximum t_{max} on curve 2 (Fig. 8.6) to continue the etching until the moment $\mathbf{n}t_{max}$ (Belyakov et al. 1977).

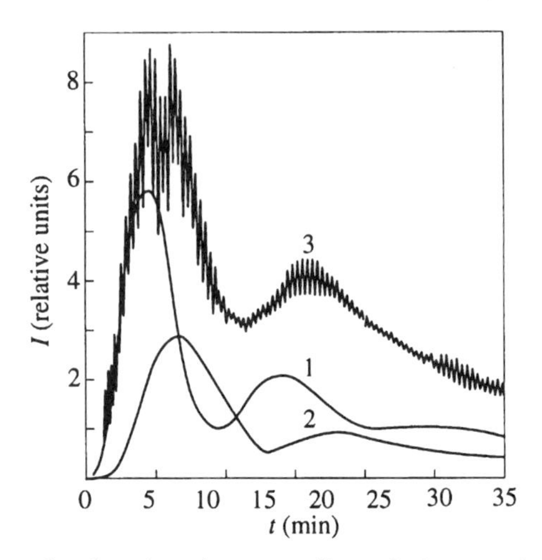

Figure 8.7. Example showing the recording of changes in diffracted light intensity during the growth of the diffraction grating: (1) Diffraction maximum of the first order; (2) diffraction maximum of the second order; (3) the interference sum of the first and second orders.

The discrepancy that appears between the arithmetic sum of the intensities of the mth and $m + 1$ order of diffraction and their interference sum, recorded in the process of etching, yields additional information on the amplitude and phase components of the grating (i.e., purely phase grating in addition to the relief) and on the deviation of the shape of relief from sinusoidal. The fluctuations of the interference sum of two orders of diffraction (Fig. 8.7, curve 3) indicate variations of phase shift in the two beams, forming the interference field or the displacement of the diffraction grating relative to this field. The causes of such phase shifts may be vibrations of the optical setup, variation of air refraction index resulting from local changes of temperature, or variations of the etchant refraction index as a result of diffusion of the reaction products.

Let us now discuss the maximum diffraction efficiency of the fabricated diffraction gratings. The optimum value of the diffraction efficiency depends on a particular applications of these gratings. For the nondistorted reproduction of holograms the diffraction efficiency must not exceed 1 percent (as long as the intensity of the second order of diffraction is negligibly small). However, for the distributed feedback semiconductor lasers, the diffraction efficiency must be as large as possible. Since the maximum value of the squared Bessel function of the first kind for the first order ($m = 1$) is equal to 0.339 for $\Gamma = 1.83$, the maximum diffraction efficiency that can be achieved for a sinusoidal relief is 33.9 percent. When a He-Ne laser with wavelength 0.63 μm is used, the maximum value of the diffraction efficiency is reached at the depth of 0.18 μm.

8.2.3. Etchant Composition

Let us consider how the composition for the precision photochemical etching of GaAs was worked out in practice (Belyakov et al. 1976b). Since the gallium arsenide is a sufficiently wide-gap material, the possibility of a corrosion or solvation mechanism of etching had to be considered. The fractional ionicity of GaAs is not large (about 35 percent; see Suchet and Bailly 1965), which points out to the corrosion mechanism of etching. GaAs has a stationary redox potential in acid solutions of about 0.25 V as opposed to SCE (Haisty 1961). Therefore GaAs is easily etched in oxidant solutions. When selecting oxidants, the colored solutions such as sulfuric acid solutions of $K_2Cr_2O_7$ and solutions of Br_2 (which give rather good results for diffraction gratings with large periods) were excluded. Using H_2O_2 as an oxidant made it possible to obtain gratings with a relief period of about 1 μm, but the evolution of oxygen during the decomposition of hydrogen peroxide prevented a further decrease of the period.

Eventually a strong oxidant—potassium persulfate $K_2S_2O_8$ in the solution of sulfuric acid—was chosen as a GaAs etchant. That solution is transparent in the visible and near UV regions of the spectrum. The product of its reduction—ions SO_4^{2-}—practically do not change the etchant properties. The optimal concentration of persulfate lies within 0.1–0.2 N, depending on the resistivity of GaAs. The concentration of sulfur acid was chosen large enough (10 N) to suppress the formation of the native GaAs oxides, mainly Ga_2O_3, and also to increase the refraction index of the etchant.

The reaction between GaAs and potassium persulfate proceeds in accordance with the sum

$$2GaAs + 8K_2S_2O_8 + 8H_2O = 2H_3AsO_4 + Ga_2(SO_4)_3 + 8K_2SO_4 + 5H_2SO_4. \quad (8.3)$$

The reduction of the ion $S_2O_8^{2-}$ has two steps (Memming 1969):

$$S_2O_8^{2-} + e^- \Rightarrow SO_4^{2-} + SO_4^{\dot{-}}, \quad (8.4a)$$

$$SO_4^{\dot{-}} + e^- \Rightarrow SO_4^{2-}. \quad (8.4b)$$

Since $SO_4^{\dot{-}}$ ion is a very strong oxidant, the second-step electrons come from the valence band, which means that hole injection takes place.

Since the first stage of the reduction of persulfate is photosensitive, the injection of holes during the second stage leads to a certain increase of the dark rate of etching. The quantum yield of the whole process amounts to 1.8, which corresponds to a two-stage etching mechanism. Using this solution, diffraction gratings with a period of about 0.2 μm and a depth of up 900 Å were fabricated at the Ioffe Institute directly on the heteroepitaxial structures GaAlAs-GaAs (Belyakov et al. 1974b). They were used for the first DFB semiconductor lasers with radiation output via the diffraction gratings.

Table 8.1 gives information about the composition of some etchants developed at the Ioffe Institute by the authors of this chapter.

8.2.4. Obtained Results

In most cases the photochemical etching is used for manufacturing holographic gratings. Below we will list a variety of n-type and p-type semiconductor materials for which special methods of diffraction grating fabrication have been developed at the Ioffe Institute. These are single crystals and monocrystalline films: Si, Ge, CdS, CdSe, CdTe, $CdGeP_2$, GaAs, GaAlAs, GaInAs, GaInAsP, GaP, GaSb, InP, and PbS; and polycrystalline and amorphous

TABLE 8.1. Examples of Composition of Etchants.

Semiconductor, Conductivity Type, Carrier Concentration, cm^{-3}	Period of the Grating, μm	Main Components of Etchant	Mechanism of Etching
Ge monocryst. n, 10^{19}	1.0	$CuSO_4$	photovolt.
Si monocryst. n, p, 10^{15}	1.0	$(NH_4)_2S_2O_8 + CrO_3 + HF$	corrosion
Se + Te polycryst. films	3.0	$K_2Cr_2O_7 + H_2SO_4$	corrosion
AsSe amorphous films	1.0	NaOH + HAc	solvat.
Same	1.0	$K_2Cr_2O_7 + H_2SO_4$	corrosion
CdS monocrystal, n	1.0	HCl	solvat.
CdSe monocrystal, n	0.3	HCl	solvat.
CdTe polycryst. films	3.0	$FeCl_3$ + HCl	corrosion
Cu_2O films, p	1.1	HOx + NaOx	solvat.
GaAs monocr. n, p, 10^{18}	0.2	$K_2S_2O_8 + H_2SO_4$	corrosion
Same	5.0	$FeCl_3$ + HCl	photovolt.
GaInAsP epitax. films	0.5	$K_2S_2O_8 + HNO_3$	corrosion
GaP monocryst. n, 10^{17}	0.25	$NOCl + HNO_3$	corrosion
GaSe polycryst. films	3.0	$K_2Cr_2O_7 + H_2SO_4$	corrosion
InAs monocrystal	3.0	$K_2Cr_2O_7 + H_2SO_4$	corrosion
InP monocryst. n, 10^{18}	2.5	$K_2S_2O_8 + HNO_3 + SnCl_4$	corrosion

Note: Ac = acetate; Ox = oxalate.

layers and films: AgCl, As_2Se_3, AsSeTe, As_2Te_3, Cu_2O, Se, and SeTe. We will also mention the original methods of photoetching CdSe (Sterligov and Tyagai 1975), GeO (Vlasenko et al. 1978), and compensated GaAs (Osgood et al. 1982).

For the minimum grating period the majority of researchers give 0.18–0.25 μm (Alferov et al. 1976a; Matz and Zirrgiebe 1988). This limit depends not on the etching method but on the wavelength of the light source. The best result, known to us, with a period of about 0.17 μm was obtained for CdS (Sterligov et al. 1976) and for GaAs (Podlesnik et al. 1983). The latter result was obtained using a UV laser. In this respect the results of the photochemical etching are close to those obtained by the photoelectrochemical method.

The maximum depth of diffraction gratings usually is on the order of 0.3–0.5 of their period, in agreement with the calculated data (Section 8.1.4). However, this dependence is not valid for long grating periods. At the Ioffe Institute gratings whose depth even exceeded their period were fabricated. However, rather than sinusoidal, their profile was closer to a triangular or rectangular shape.

The grating quality can be characterized by the diffraction efficiency and by the level of diffuse light scattering. The best of the sinusoidal gratings fabricated at the Ioffe Institute (on GaAs) had diffraction efficiency of about 30 percent, which is close to 90 percent of the theoretically achievable values (33.9 percent), and as far as we know, that exceeds the results obtained by other authors.

The scattering of light is mainly caused by pronounced surface imperfections appearing after a long exposition. For the optimally selected etchants, the total level of diffuse light scattering, according to our data, does not exceed 0.1 percent.

The geometric sizes of the gratings, fabricated by photochemical etching, are limited by the optic setup. Apart from a considerable widening of light beams, one must provide a uniform illumination of the semiconductor area. We fabricated diffraction gratings on the GaAs wafers with a diameter of ~ 30 mm and on the As_2Se_3 vitreous layers whose sizes exceeded 70×70 mm^2. (The publications by other authors do not, as a rule, mention the grating sizes.)

We should stress that these diffraction gratings had a *sinusoidal* profile. Such gratings cannot be fabricated using photoresistive masks. In this respect photochemical etching is unique.

8.3. PHOTOCHEMICAL ETCHING OF METALS

Even though the chemical etching of metals is not photosensitive, the experiments at the Ioffe Institute have shown that a surface relief corresponding to the light distribution can be also obtained on metals (Belyakov et al. 1983). We started from the idea that the semiconductor properties of films, formed at the interface of the metal and the etchant, could play a decisive role.

When metals are etched, an oxide or salt film is often formed on their surfaces. The thickness of this film eventually determines the rate of the metal dissolution. This thickness depends on the relation between the formation and dissolution rates. If the film has semiconductor properties (which is rather characteristic of many oxide films), then the dissolution rate depends on the film illumination. Hence illumination affects the thickness of the film and, eventually, the metal dissolution rate.

Copper was chosen as the main object of investigation. When copper is etched, oxide films can form on its surface. We have already shown a possibility of photosensitive dissolution of one of the copper oxides—Cu_2O (Goryachev and Paritskii 1974; Belyakov et al. 1979c). The studies of the chemical composition and the polarization, spectral, and capacitance- and lux-ampere characteristics of the copper oxide films showed that they can be a light-sensitive component in the metal-film-solution system (Belyakov et al. 1984). Certain regularities of the formation and dissolution of such films, as well as their semiconductor properties, have also been studied. Such investigations showed that the film *formation* rate on copper (which is in contact with the oxidant) decreases with time t, as $t^{0.5}$; namely it is determined by the diffusion processes in the bulk of the film and does not depend on the illumination. Its *dissolution* rate in the solutions of some complexing agents is constant in time but depends on the illumination. Hence there must be a local dynamic balance between the formation and dissolution rates at any part of the surface. This balance is determined by the local illumination. Since the film formation takes place due to the "destruction" of the equivalent amount of the metal, the metal dissolution rate is determined by its local illumination. For metal etching we developed special solutions that were both oxidants for the metal and etchants for the oxide film. As a result these were produced holographic gratings on copper with a period of 1 μm, the depth of the sinusoidal profile of up to 0.3 μm (in case of 4 μm period) and the diffraction efficiency up to 17.5 percent (Belyakov et al. 1983). The gratings were also produced on nickel and silver. These results, according to our data, are much better than the state-of-art results obtained by chemical methods. Only on aluminum films the gratings with a period of 1.5 μm were obtained using the thermochemical properties of oxide aluminum films (Tsao and Ehrlich 1983).

8.4. APPLICATIONS OF PHOTOCHEMICAL ETCHING

8.4.1. Applications in Optoelectronics

A periodic structure can substitute a Fabri-Perot resonator in the semiconductor laser. In 1971 Kogelnik and Shank demonstrated a laser with an active medium whose refraction index changes periodically. Later such lasers were called *lasers with a distributed feedback* (or DFB lasers). Also in 1971 a new type of laser with a diffraction grating on the surface of the active heterostructure layer was developed at the Ioffe Institute. Such a grating diminishes the radiation divergence and at the same time uses the DFB effect. Kazarinov and Suris (1972) have developed a theory of such a laser. Photochemical etching was used to fabricate these lasers. A diffraction grating with a period of about 0.22 μm on the waveguide surface made it possible to fabricate a semiconductor laser with a small divergence (about 30 angular minutes) with the light output through the plane parallel to the active layer (Alferov et al. 1975). It was also shown that the grating produced by this method did not lead to appreciable losses.

Soon the DFB lasers were fabricated with a light output through the diffraction grating with a depth of about 900 Å (Alferov et al. 1976b). Later a similar work was carried on in other countries (Lum et al. 1985; Aoyagi et al. 1985). Related work on heterostructure wave guides was reported for GaAlAs system (Alferov et al. 1976c; Dias et al. 1976; Kalandarishvili et al. 1981), for the CdSSe system (Buchaidze et al. 1982), for the GaPAs system, and for the GaN system (Andreev et al. 1978).

Photochemical etching has proved to be an attractive method for manufacturing optical holographic filters for the long wavelength lasers when it is necessary to modulate the grating relief with another spatial frequency (Grebel and Pien 1992).

8.4.2. Optical Properties of Metal-Coated Microprofiled Surfaces

Precision photochemical etching was applied to fabrication of Schottky diodes (metal-semiconductor structures), MIS- (metal-isolator-semiconductor), and MOM- (metal-oxide-metal) structures with interface gratings.

The excitation of surface electromagnetic waves coupled with plasmons (surface plasmon-polaritons) was demonstrated when light interacted with corrugated metal surfaces (Agranovich and Mills 1982). The above-mentioned structures exhibited a photocurrent resonance when the wavelength, the angle of incidence, and polarization of light met the coupling conditions (Belyakov et al. 1985). These resonance phenomena were used in spectral and polarization-sensitive photodetectors (Belyakov et al. 1990, 1992; Sreseli et al. 1992). For instance, the selective photodetector based on Ag/n–GaAs had a spectral half width of 4 nm for the wavelength of 0.6328 μm. The ratio of the photocurrent in the resonance to the photocurrent far from the resonance was close to 25. With the shift to longer wavelengths, the resonance parameters improve.

The photoelectric method of the diagnostics and investigation of the surface plasmon-polaritons has been developed for the modified Schottky diodes with sinusoidally corrugated interfaces (Belyakov et al. 1991a). This method revealed many peculiarities of those waves, in particular, the "slow" plasmon-polaritons at the metal-semiconductor interface (Belyakov et al. 1989a, 1989b). It was also discovered that the presence of the diffraction grating on the metal surface eliminates the restrictions for the light polarization (Belyakov et al. 1987a).

The coupling of surface plasmon-polaritons was investigated with a simultaneous excitation of such two waves as those propagating on the same interface in different directions (Belyakov et al. 1991b) and those running on the both corrugated surfaces of the metal layer in the same direction (Sreseli et al. 1991a, 1991b).

The diffraction gratings on the semiconductor surface fabricated by precision photochemical etching were also used for studying the spectrum splitting of the surface phonon-polaritons (Goryachev et al. 1984), the resonance suppression of the mirror reflection from the deep semiconductor gratings (Belyakov et al. 1987b), and the resonance diffraction of light in the exciton region of the semiconductor absorption spectrum (Goryachev et al. 1990)

A regular microprofiling of the surface changes not only optical but also other physical properties of the surface region of the semiconductors. For example, the potential relief related to the geometrical relief, appears on the corrugated surface of the semiconductor (Belyakov et al. 1981, 1982)

8.4.3. Other Applications of Precision Photochemical Etching

It goes without saying that the precision photochemical etching can be used not only for the diffraction grating fabrication or recording other holographic images but also for the "drawing" on a semiconductor surface by a light beam, for instance, for etching deep grooves in GaAs (Podlesnik et al. 1986) or for opening windows in heterostructures (Moutonnet 1987).

Precision photochemical etching is also a very good method of investigating *in situ* both the photochemical etching processes and the properties of the semiconductor-electrolyte interface. The advantage of this method of studying the photostimulated processes is high sensitivity (the formation of relief can be fixed at its average depth of 50 Å) and the possibility to observe the integral effect on the whole area of the specimen, so the effect of local defects on the semiconductor is averaged out.

It was explained, for example, why photochemical etching of the semiconductor vitreous films takes place in a wide spectral range (Belyakov et al. 1977) while the "dry" recording of the optic information on the same materials can be observed only in a very narrow region of the spectrum. Another example is the chemical polishing of semiconductor surfaces. To this end, first a diffraction grating with the given period and the given depth is formed on a semiconductor surface. Then the sample is subjected to a chemical polishing. During chemical polishing unusual dependences of the relief depth's change on its spatial frequency were discovered (Goryachev et al. 1983).

8.5. CONCLUSION

This chapter has described the basic principles of precision maskless photochemical etching of semiconductors. The method has proved to be suitable for manufacturing reflective diffraction gratings with a sinusoidal profile. Such gratings are employed in optoelectronic devices where the preservation of the ideal crystal structure of the semiconductor is extremely important. This method is comparable to photoresistive etching of semiconductors. The unique advantages of the method are its ability to adjust the shape of the relief by illumination and the ability to control the etching process. Also photochemical etching requires a smaller number of technological steps compared with other techniques.

REFERENCES

Agranovich, V. M. and Mills, D. L., eds. (1982). *Surface Polaritons*. North-Holland, Amsterdam.

Alferov, Zh. I., Gurevich, S. A., Kazarinov, R. F., et al. (1975). *Sov. Phys. Semicond.* 8, 1321–1322.

Alferov, Zh. I., Goryachev, D. N., Gurevich, S. A., et al. (1976a). *Sov. Phys. Tech. Phys.* 21, 857–859.

Alferov, Zh. I., Gurevich, S. A., Klepikova, N. V., et al. (1976b). *Sov. Tech. Phys. Lett.* 2, 245–250.

Alferov, Zh. I., Gurevich, S. A., Klepikova, N. V., et al. (1976c). *Sov. Phys. Tech. Phys.* 21, 320–322.

Andreev, V. M., Bykovsky, Ju. A., Vigdorovich, E. N., et al. (1978). *Sov. J. Quantum Electron.* 5, 135–138.

Aoyagi, Y., Masuda, S., Doi, A., et al. (1985). *Jap. J. Appl. Phys.* 24, L294–L296.

Belyakov, L. V., Goryachev, D. N., Ostrovskii, Yu. I., and Paritskii, L. G. (1974a). *Zh. Nauchn. Prikl. Fotogr. Kinematogr.* (USSR) 19, 54–56.

Belyakov, L. V., Goryachev, D. N., Mizerov, M. N., and Portnoi, E. L. (1974b). *Sov. Phys. Tech. Phys.* 19, 837–838.

Belyakov, L. V., Goryachev, D. N., Paritskii, L. G., Ryvkin, S. M., and Sreseli, O. M. (1976a). *Sov. Phys. Semicond.* 10, 678–681.

Belyakov, L. V., Goryachev, D. N., Paritskii, L. G., and Sreseli, O. M. (1976b). *Fiz. i. Tekhn. poluprovodnikov* (USSR) 10, 1603.

Belyakov, L. V., Goryachev, D. N., and Sreseli, O. M. (1977). *Sov. Tech. Phys. Lett.* 3, 377–378.

Belyakov, L. V., Goryachev, D. N., Ryvkin, S. M., and Sreseli, O. M. (1978). *Sov. Phys. Semicond.* 12, 1134–1137.

Belyakov, L. V., Goryachev, D. N., Ryvkin, S. M., Sreseli, O. M., and Suris, R. A. (1979a). *Sov. Phys. Semicond.* 13, 1270–1273.

Belyakov, L. V., Goryachev, D. N., and Sreseli, O. M. (1979b). *Sov. Phys. Tech. Phys.* 24, 511–512.

Belyakov, L. V., Goryachev, D. N., and Sreseli, O. M. (1979c). *Sov. Tech. Phys. Lett.* 5, 406–407.

Belyakov, L. V., Goryachev, D. N., and Sreseli, O. M. (1981). *Sov. Phys. Semicond.* 15, 1205–1206.

Belyakov, L. V., Goryachev, D. N., Sreseli, O. M., and Suris, R. A. (1982). In *Trudy Vses. Konf. po fizike poluprovodnikov*, vol. 2, pp. 79–80. ELM, Baku, USSR.

Belyakov, L. V., Goryachev, D. N., and Sreseli, O. M. (1983). *Sov. Tech. Phys. Lett.* 9, 204–205.

Belyakov, L. V., Goryachev, D. N., and Sreseli, O. M. (1984). *Sov. Phys. Semicond.* 18, 470–471.

Belyakov, L. V., Goryachev, D. N., Sreseli, O. M., and Yaroshetskii, I. D. (1985). *Sov. Tech. Phys. Lett.* 11, 481–482.

Belyakov, L. V., Goryachev, D. N., Emel'yanov, V. I., et al. (1987a). *Sov. Tech. Phys. Lett.* 13, 228–290.

Belyakov, L. V., Goryachev, D. N., Sreseli, O. M., and Yaroshetskii, I. D. (1987b). *Sov. Tech. Phys. Lett.* 13, 353–354.

Belyakov, L. V., Goryachev, D. N., Rumyantsev, B. L., Sreseli, O. M., and Yaroshetskii, I. D. (1989a). *Sov. Phys. Semicond.* 23, 288–291.

Belyakov, L. V., Goryachev, D. N., Makarova, T. L., et al. (1989b). *Sov. Phys. Semicond.* 23, 1217–1220.

Belyakov, L. V., Goryachev, D. N., Rumyantsev, B. L., Sreseli, O. M., and Yaroshetskii, I. D. (1990). *Sov. Tech. Phys. Lett.* 16, 235–236.

Belyakov, L. V., Goryachev, D. N., Rumyantsev, B. L., Sreseli, O. M., and Yaroshetskii, I. D. (1991a). *Bull. Acad. Sci. USSR, Phys. Ser.* 7, 1290–1295.

Belyakov, L. V., Vaksman, V. I., Goryachev, D. N., et al. (1991b). *Sov. Phys. Tech. Phys.* 36, 645–648.

Belyakov, L. V., Goryachev, D. N., Rumyantsev, B. L., Sreseli, O. M., and Yaroshetskii, I. D. (1992). *Electrosvyaz* (*USSR*) (11), 20–21.

Buchaidze, Z. E., Vasilishcheva, I. V., Morozov, V. N., et al. (1982). *Sov. J. Quantum* 12, 1514–1516.

Dalisa, A. and DeBitetto, D. (1970a). *Appl. Optics* 11, 2007–2015.

Dalisa, A., Zwicker, W., DeBitetto, D., and Harnack, H. (1970b). *Appl. Phys. Lett.* 17, 208–210.

Dias, P., Klepikova, N. V., Mizerov, M. N., et al. (1976). *Pis'ma v Zh. Tekh. Fiz.* (USSR) 2, 347–351.

Egorov, B. V., Karpov, S. Ju., Mizerov, M. N., Portnoi, E. L., and Smirnitskii, V. B. (1984). *Sov. Phys. Tech. Phys.* 29, 1145–1149.

Flanders, D. C., Hawryluk, A. M., and Smith, H. I. (1979). *J. Vac. Sci. Technol.* 16, 1949–1952.

Golubenko, G. A., Prokhorov, A. M., Sychugov, V. A., and Tulajkova, T. V. (1985). *Phys. Chem. Mechan. Surf.* (1), 88–92.

Goryachev, D. N., Paritskii, L. G., and Ryvkin, S. M. (1970a). *Fiz. i Tekhn. poluprovodnikov* (USSR) 4, 1582–1583.

Goryachev, D. N., Paritskii, L. G., and Ryvkin, S. M. (1970b). *Fiz. i Tekhn. poluprovodnikov* (USSR) 4, 1580–1581.

Goryachev, D. N. and Paritskii, L. G. (1974). *Sov. Phys. Semicond.* 7, 972.

Goryachev, D. N., Paritskii, L. G., and Sreseli, O. M. (1975). *Fiz. i Tekhn. poluprovodnikov* (USSR) 9, 1222.

Goryachev, D. N., Belyakov, L. V., and Sreseli, O. M. (1983). *Sov. Electrochem.* 19, 720–722.

Goryachev, D. N., Dmitruk, N. K., Kamuz, A. M., and Litovchenko, V. G. (1984). *Poverkhnost* (USSR) (2), 44–47.

Goryachev, D. N., Ismailov, K. A., Kosobukin, V. F., Sazhin, M. I., and Sel'kin, A. V. (1990). *Opt. Spectroscopy* 69, 1201–1203.

Grebel, H. and Pien, P. (1992). *J. Appl. Phys.* 71, 2428–2432.

Gurevich, Yu. Ya. and Pleskov, Yu. V. (1983). *Fotoelektrokhimiya poluprovodnikov*, pp. 252–261. Nauka, Moscow.

Haisty, R. (1961). *J. Electrochem. Soc.* 108, 790–794.

Kalandarishvili, K. G., Mizerov, M. N., Nikishkin, S. A., et al. (1981). *Sov. Tech. Phys. Lett.* 7, 641–642.

Kazarinov, R. F. and Suris, R. A. (1972). *Sov. Phys. Semicond.* 6, 1184–1189.

Kogelnik, H. and Shank, C. V. (1971). *Appl. Phys. Lett.* 18, 152–154.

Kuhn-Kuhnenfeld, F. (1972). *J. Electrochem. Soc.* 119, 1063–1068.

Lum, R. M., Glass, A. M., Ostermayer, F. W., et al. (1985). *J. Appl. Phys.* 57, 39–44.

Matz, R. and Zirrgiebe, J. (1988). *J. Appl. Phys.* 64, 3402–3406.

Memming, R. (1969). *J. Electrochem. Soc.* 116, 785–790.

Moutonnet, D. (1987). *J. Appl. Phys.* B42, 221–223.

Myamlin, V. A. and Pleskov, Yu. V. (1967). *Electrochemistry of Semiconductors*. Plenum Press, New York.

Osgood, R. V., Sanchez-Ribo, A., Ehrlich, D. J., and Daneu, V. (1982). *Appl. Phys. Lett.* 40, 391–393.

Pleskov, Yu. V. (1960). *Doklady Akad. Nauk SSSR* (USSR) 132, 1360–1363.

Podlesnik, D. V., Gilgen, H. H., Osgood, R. V., and Sanchez, A. (1983). *Appl. Phys. Lett.* 42, 1083–1085.

Podlesnik, D. V., Gilgen, H. H., and Osgood, R. M. (1986). *Appl. Phys. Lett.* 48, 496–498.

Rytz-Froidevaux, Y. and Salathe, R. P. (1985). *J. Appl. Phys.* A37, 121–138.

Sreseli, O. M., Belyakov, L. V., Goryachev, D. N., Rumyantsev, B. L., and Yaroshetskii, I. D. (1991a). In *Proc. SPIE* (SU) 1440, 326–331.

Sreseli, O. M., Belyakov, L. V., Goryachev, D. N., Rumyantsev, B. L., and Yaroshetskii, I. D. (1991b). In *Proc. SPIE* (USA) 1545, 149–158.

Sreseli, O. M., Belyakov, L. V., Goryachev, D. N., Rumyantsev, B. L., and Yaroshetskii, I. D. (1992). In *Proc. SPIE* (USA) 1735, 37–42.

Sterligov, V. A. and Tyagai, V. A. (1975). *Pis'ma v Zh. Tekh. Fiz.* (USSR) 1, 704–708.

Sterligov, V. A. Kolbasov, G. Ya., and Tyagai, V. A. (1976). *Pis'ma v Zh. Tekh. Fiz.* (USSR) 2, 437–440.

Suchet, J. P. and Bailly, F. (1965). *Ann. Chem.* 10, 517–520.

Sychugov, V. A. and Tulajkova, T. V. (1984). *Sov. J. Quantum Electr.* 14, 301–302.

Tsao, J. Y. and Ehrlich, D. J. (1983). *Appl. Phys. Lett.* 43, 146–148.

Tsukada, N., Sugata, S., Saitoh, H., and Mita, Y. (1983). *Appl. Phys. Lett.* 43, 189–191.

Vlasenko, N. A., Nazarenkov, F. A., Sterligov, V. A., and Tyagai, V. A. (1978). *Sov. Phys. Tech. Phys.* 4, 537–538.

INDEX